Parallel Programming
with Co-arrays

Parallel Programming with Co-arrays

Robert W. Numrich

CRC Press
Taylor & Francis Group
Boca Raton London New York

CRC Press is an imprint of the
Taylor & Francis Group, an **informa** business
A CHAPMAN & HALL BOOK

CRC Press
Taylor & Francis Group
6000 Broken Sound Parkway NW, Suite 300
Boca Raton, FL 33487-2742

Printed on acid-free paper
Version Date: 20180716
Printed by CPI Group (UK) Ltd, Croydon CR0 4YY

International Standard Book Number-13: 978-1-4398-4004-7 (Hardback)
International Standard Book Number-13: 978-0-4297-9327-1 (Paperback)

Library of Congress Cataloging-in-Publication Data

Names: Numrich, Robert W., author.
Title: Parallel programming with co-arrays / Robert W. Numrich.
Description: First edition. | Boca Raton, FL : CRC Press/Taylor & Francis
Group, 2018. | Series: Chapman & Hall/CRC computational science ; 33 |
Includes bibliographical references and index.
Identifiers: LCCN 2018024855 | ISBN 9781439840047 (hardback : acid-free paper)
Subjects: LCSH: Parallel processing (Electronic computers)
Classification: LCC QA76.642 .N86 2018 | DDC 004/.35--dc23
LC record available at https://lccn.loc.gov/2018024855

Visit the Taylor & Francis Web site at
http://www.taylorandfrancis.com

and the CRC Press Web site at
http://www.crcpress.com

Contents

Preface

This book describes basic parallel algorithms encountered in scientific and engineering applications. It implements them in modern Fortran using the co-array programming model and it analyzes their performance.

The book's intended audience includes graduate and advanced undergraduate students who want an introduction to basic techniques. It also includes experienced researchers who want to understand the relationships among seemingly disparate techniques. The book assumes that the reader knows the modern Fortran language, basic linear algebra, and basic techniques for solving partial differential equations.

The book contains many code examples. Each example has been tested, but there is no warranty that the code is free of errors. Nor is there any warrant that a particular example represents the best implementation. Reading code without writing code, however, is not enough. The only way to learn parallel algorithms is to write code and to test it. Modifying the sample codes is a good way to experiment with alternative implementations.

The entries in the Bibliography provide a guide to the development of parallel programming techniques over the last few decades. It is not exhaustive, and it does not provide an accurate historical record of how they evolved or who first formulated them. They simply provide a guide to more detailed discussions of topics covered in the book.

Many thanks to John Reid whose encouragement and support helped make this book possible. He worked through many of the technical details of the co-array model and shepherded it through the rigorous standardization process to make it part of the Fortran language. He read several versions of the manuscript, and I greatly appreciate his suggestions for improvement.

Robert Numrich
Minneapolis, Minnesota

Chapter 1

Prologue

This book describes a set of fundamental techniques for writing parallel application codes. These techniques form the basis for parallel algorithms frequently used in scientific and engineering applications. Parallel computing by itself is a very large topic as is scientific computing by itself. The book makes no claim of comprehensive coverage of either topic, just a basic outline of how to write parallel code for scientific applications.

All the examples in the book employ two fundamental techniques that are part of every parallel programming model in one form or another:

- data decomposition

- execution control

The programmer must master these two techniques and may find them the hardest part of designing a parallel application. The book applies these two fundamental techniques to five fundamental algorithms:

- matrix-vector multiplication

- matrix factorization

- matrix transposition

- collective operations

- halo exchanges

It is not a complete list, but it is a list that every parallel code developer must understand.

The book describes these techniques in terms of partition operators. The programmer frequently encounters partition operators, either explicitly or implicitly, in scientific application codes, and the techniques needed for new codes are often variations of techniques encountered in previous codes. The specific form of the partition operators becomes progressively more complicated as the examples become more complicated. The book's goal is to show that all its examples fit into a single unified framework.

The book encourages the use of Fortran as a modern object-oriented language. Parallel programming is an exercise in transcribing mathematical definitions for partition operators into small functions associated with Fortran objects. The exchange of one data distribution for another is the exchange of one set of functions with another set. This technique follows one of the fundamental principles of object-oriented design.

This book demonstrates that the programmer needs to learn just a handful of techniques that are variations on a common theme. As always, however, the only way to learn to write parallel code is to write parallel code. And to test it.

The interplay between two sets of indices makes the programmer's job difficult. One set describes the decomposition of a data structure. The other set describes how the first set is distributed across a parallel computer. Perhaps the statement attributed to Kronecker,

"God created the integers, all else is the work of man."

applies just as well to the whole of parallel programming as it does to the whole of mathematics [47].

Chapter 2

The Co-array Programming Model

This chapter explains the essential features of the co-array programming model. Programmers use co-array syntax to move data objects from one local memory to another local memory. To ensure that the values moved from one memory to another are the correct values, programmers insert explicit synchronization statements to control execution order.

The co-array programming model follows the Single-Program-Multiple-Data (SPMD) execution model. The run-time system creates multiple copies of the same program and executes the statements in each copy asynchronously in parallel. The co-array model calls each copy of the program an image. The run-time system assigns a unique image index to each copy, and the number of images remains fixed throughout the execution of a co-array program [56] [70].

The book uses imprecise expressions such as, "the image executes a statement" to mean that "the run-time system executes a statement within a copy of the program assigned to the image." The second expression is more precise than the first because an image is just a sequence of statements that can do nothing on their own. The minor imprecision in the first expression, however, is worth the gain in readability.

2.1 A co-array program

Before looking at specific parallel algorithms, the book first describes the basic ideas behind the co-array programming model by examining the program shown in Listing 2.1. The run-time system creates multiple copies of this program, called images, assigns each image to physical hardware, and allocates local memory for each image with affinity to the hardware. The run-time system executes the statements independently for each image according to the normal rules of the Fortran language.

Because each image is a copy of the same program, variables declared in the program have the same names across all images. One variable in the program, the integer variable x[:], is an allocatable co-array variable declared with one unspecified co-dimension in square brackets. In addition to integer variables, complex, character, logical or user-defined variables can be declared as co-array variables. A co-array variable has the special property that it is visible across images. At any time during execution, an image can reference both the value of its own variable and the value of the variable assigned to any other image. The other four variables, me,p,y,you, declared in the program are normal variables. Their values are visible only within each image, and they cannot be referenced by a remote image.

Listing 2.1: A co-array program.

```
program First
   implicit none
   integer,allocatable :: x[:]  !----co-array variable--!
   integer :: me,p,y,you        !----normal variables---!

     p   = num_images()         !----begin segment 1----!
     me  = this_image()         !                       !
     allocate(x[*])             !----end    segment 1----!

     you = me+1                 !----begin segment 2----!
     if(me == p) you = 1        !                       !
     x = me                     !                       !
     sync all                   !----end    segment 2----!

     y = x[you]                 !----begin segment 3----!
     write(*,"('me:    ',i5,' my pal:    ',i5))") me, y  !
                                !                       !
     deallocate(x)             !----end    segment 3----!

                               !----begin segment 4----!
           :                   !                       !
           :                   !                       !
   end program First           !----end    segment 4----!
```

The programmer needs to know the number of images and the value of the local image index at run-time. Each image obtains these values by invoking two intrinsic functions added to the language to support the co-array model as replicated in Listing 2.2.

Listing 2.2: Intrinsic functions.

```
     p   = num_images()
     me  = this_image()
```

The first function returns the number of images, and the second function returns the image index of the image that invokes it. The image index has a value between one and the number of images and uniquely identifies the image. In all the examples that follow, the book uses the symbol p for the number of images and the symbol me for the value of the local image index.

Listing 2.3: Allocating a co-array variable with default co-bounds.

```
allocate(x[*])
```

The allocation statement, replicated in Listing 2.3, sets the upper and lower co-bounds for the co-dimension. Specifying the co-bounds with an asterisk in square brackets follows the Fortran convention for declaring a normal variable with an assumed size. This convention allows the programmer to write code that can run on different numbers of images without changing the source code, without re-compiling and without re-linking. The implied lower co-bound is one and the implied upper co-bound equals the number of images at run-time.

The programmer can override the default co-bound values as described in Section A.3. If the lower co-bound is zero, for example, the upper co-bound is the number of images minus one. Whatever the value for the lower co-bound, the programmer is responsible for using valid co-dimension indices. Compiler vendors may provide an option to check the run-time value of a co-dimension index, but an out-of-bound value for a co-dimension index results in undefined, often disastrous, behavior just as it does for an out-of-bound value for a normal dimension index.

In the program shown in Listing 2.1, each image picks a partner with image index one greater than its own. Since the partner's index cannot be greater than the number of images, the last image picks the first image as its partner. Each image sets the value of its local co-array variable to its own image index as shown in Listing 2.4.

Listing 2.4: Reference to a local co-array variable.

```
x=me
```

A reference to a co-array variable without a co-dimension index is equivalent to a reference with an explicit co-dimension index equal to the local image index as shown in Listing 2.5.

Listing 2.5: Alternative reference to a local co-array variable.

```
x[me]=me
```

This convention is a fundamental feature of the co-array model. The default view of memory is a local view that allows the programmer to write code without redundant syntax. It also eliminates the need for extra analysis, both at compile-time and at run-time, to recognize a purely local memory reference.

The programmer inserts execution control statements into a program to impose an execution order across images. The programmer is responsible for placing execution control statements correctly and may find this requirement the most difficult aspect of the co-array model.

The program shown in Listing 2.1 consists of four segments determined by three execution control statements plus the statement that ends the program. The first segment begins with the first executable statement and ends with the allocation statement. Every image must execute this statement because the variable being allocated is a co-array variable. There is an implied barrier that no image may cross until all have allocated the variable.

The second segment consists of all statements following the allocation statement up to and including the `sync all` statement. Without this control statement, an image might try to obtain a value from its partner before the partner has defined the value. The `sync all` statement guarantees that each image has defined its value before any image references the value. When an image executes a statement in the third segment, as shown in Listing 2.6, it obtains the value defined by its partner in the second segment.

Listing 2.6: Accessing the value of a co-array variable assigned to a remote image.

```
y = x[you]
```

The third segment ends with execution of the deallocation statement. Because the variable is a co-array variable, every image waits until all images reach this statement before deallocating the variable. Otherwise, an image might try to reference the value of a variable that no longer exists. The fourth segment consists of other statements between the deallocation statement and the end of program statement assuming there are no more control statements in the program.

Listing 2.7: Output from the program shown in Listing 2.1.

```
me:   5   my pal:   6
me:   3   my pal:   4
me:   1   my pal:   2
me:   2   my pal:   3
me:   4   my pal:   5
me:   6   my pal:   7
me:   7   my pal:   1
```

Execution of this program using seven images might result in the output shown in Listing 2.7. Each image has obtained the correct value from its partner, but output from different images appears randomly in the standard output file. The reason for this behavior is that the run-time system executes the output statement independently for each image. It is required to write a complete record for each image, with no intermixing of records from different images, and to merge records from different images into a single file. The order in which it merges the records, however, is implementation specific, and the order may vary from execution to execution even on the same system.

To guarantee order in an output file, the programmer may need to restrict execution of output statements to a single image that gathers data from other images before writing to the file in a specific order as shown, for example, in

Listing 2.8. The alternative form of the output statement avoids the need for an additional variable to hold the value obtained from the remote image.

Listing 2.8: Enforcing output order.

```
if (me == 1) then
  do q=1,p
    write(*,"('me:    ',i5,'  my pal:    ',i5))") q, x[q]
  end do
end if
```

2.2 Exercises

1. Change the example program shown in Listing 2.1 in the following ways.

   - Remove the `sync all` statement and describe what happens.
   - What happens if the variable y, just before the write statement, is replaced by the co-array variable x?
   - Change the code so that each image sends its image index to its partner.
   - Change the code so that one image broadcasts a variable to all other images.
   - Change the code so that each image obtains a variable from image one.
   - Pick one image and have it add together the values of some variable across all images.
   - Have all images add the values at the same time.
   - Have one image add the values and broadcast it to the other images.

2. In Listing 2.8, have image p perform output in reverse order.

Chapter 3

Partition Operators

The first problem a programmer faces when implementing a parallel application code is the data decomposition problem. In the co-array model, the programmer partitions data structures into pieces and assigns each piece to an image. The run-time system places each piece in the appropriate local memory and executes code independently for each image. The programmer uses co-array syntax to move data between local memories and inserts execution control statements to order statement execution among images.

Partition operators provide a mathematical description of the data decomposition process. A partition operator cuts an index set into subsets and induces a partition of a set of objects labeled by the index set, for example, a partition of the elements of a vector labeled by the element indices. The mathematical formulas that define a partition operator transcribe directly into Fortran code within the co-array model.

Index manipulations make parallel programming hard. A serial algorithm works with a set of data objects labeled by global indices. A parallel version of the same algorithm works with subsets of the data objects labeled by local indices. The map between global and local indices may be simple for simple algorithms but becomes more complicated for more complicated algorithms. Errors mapping indices, forward global-to-local and backward local-to-global, are the source of many difficulties in the design of parallel algorithms. Precise definitions for partition operators prevent some of these errors.

3.1   Uniform partitions

The fundamental object of interest in scientific computing is the finite set of ordered integers,

$$N = \{1, \ldots, n\} . \tag{3.1}$$

This global index set labels a set of objects,

$$O = \{o_1, \ldots, o_n\} \ . \tag{3.2}$$

The order relationship is the natural integer order, but for some applications a permutation may change the order relationship.

By default, the Fortran language assumes the value one for the lower bound of the index set. It allows the programmer to override this default value replacing it, for example, with zero when interfacing with other languages. Changing the default lower bound, however, usually adds little benefit for algorithm design.

Parallel applications require the programmer to partition the global index set into local index sets. If n, the size of the global index set, is a multiple of p, the number of partitions, the size of the local index set,

$$m = n/p \ , \tag{3.3}$$

is the same for each partition. The global base index for each partition has the value,

$$k_0^\alpha = (\alpha - 1)m \ , \quad \alpha = 1, \ldots, p \ , \tag{3.4}$$

and the local index set is a contiguous set of integers added to the base,

$$N^\alpha = \{k_0^\alpha + 1, \ldots, k_0^\alpha + m\} \ , \quad \alpha = 1, \ldots, p \ . \tag{3.5}$$

In scientific computing, the set of objects (3.2) is often a vector of real or complex numbers,

$$x = \begin{bmatrix} x_1 \\ \vdots \\ x_n \end{bmatrix} \ . \tag{3.6}$$

The partition of the global index set induces a partition of the vector elements,

$$\begin{bmatrix} x_1^\alpha \\ \vdots \\ x_m^\alpha \end{bmatrix} = \begin{bmatrix} x_{k_0^\alpha + 1} \\ \vdots \\ x_{k_0^\alpha + m} \end{bmatrix} \ . \tag{3.7}$$

The numerical values of the vector elements are the same, but they are labeled by the local index set,

$$L = \{1, \ldots, m\} \ , \tag{3.8}$$

rather than by the global index set.

Partitioning a vector is an operation in linear algebra,

$$x^\alpha = P^\alpha x \ , \quad \alpha = 1, \ldots, p \ . \tag{3.9}$$

The partition operator is represented by the rectangular matrix,

$$P^\alpha = \begin{bmatrix} 0 & \cdots & I^\alpha & \cdots & 0 \end{bmatrix} \ , \tag{3.10}$$

with m rows and n columns. Each zero symbol represents a zero matrix of size $m \times m$, and the symbol I^α represents the identity matrix of the same size. The partition operation defined by formula (3.9), then, is the matrix-vector multiplication,

$$
\begin{bmatrix} x_{k_0^\alpha+1} \\ \vdots \\ x_{k_0^\alpha+m} \end{bmatrix} = \begin{bmatrix} 0 & \cdots & I^\alpha & \cdots & 0 \end{bmatrix} \begin{bmatrix} x_1 \\ \vdots \\ x_{k_0^\alpha+1} \\ \vdots \\ x_{k_0^\alpha+m} \\ \vdots \\ x_n \end{bmatrix} . \tag{3.11}
$$

Partition operators define relationships among three sets of indices: the global index set, the local index set, and the image index set. When the number of partitions equals the number of images, the partition index and the image index are the same. A vector element x_k, with global index k, maps to image index α according to the formula,

$$
\alpha = \left\lfloor \frac{k-1}{m} \right\rfloor + 1 , \tag{3.12}
$$

where the notation $\lfloor \cdot \rfloor$ denotes the floor function. The same element considered as a local element assigned to image α is labeled with a local index i according to the formula,

$$
i = k - k_0^\alpha . \tag{3.13}
$$

On the other hand, a local index i assigned to partition α maps back to the global index k according to the formula,

$$
k = k_0^\alpha + i , \quad i = 1, \ldots, m . \tag{3.14}
$$

The partition operator, therefore, defines a forward map from one global index to two local indices, a partition index and a local index,

$$
k \to (\alpha, i) , \tag{3.15}
$$

according to furmulas (3.12) and (3.13), and a backward map from the two local indices to the global index,

$$
(\alpha, i) \to k , \tag{3.16}
$$

according to formula (3.14). Implementing a parallel algorithm requires precise application of these maps.

3.2   Non-uniform partitions

When the size of the global index set is not a multiple of the number of partitions, the partition sizes are non-uniform and the index maps change. There is more than one way to define these maps, but partition sizes should be as close to uniform as possible, and the maps between indices should be natural extensions of the maps for a uniform partition.

The formulas for the new maps depend on the remainder,

$$r = n \bmod p \, , \tag{3.17}$$

and on the quantity,

$$m = \lceil n/p \rceil \, , \tag{3.18}$$

where $\lceil \cdot \rceil$ is the ceiling function. The size assigned to partition α obeys the formula,

$$m^{\alpha} = \begin{cases} m & r = 0 \quad \alpha = 1, \dots, p \\ m & r > 0 \quad \alpha = 1, \dots, r \\ m - 1 & r > 0 \quad \alpha = r+1, \dots, p \end{cases} \, . \tag{3.19}$$

When the remainder is zero, the partition is uniform and the size is the same for all images. When the remainder is not zero, the partition size for images with indices less than or equal to the remainder equals the ceiling value. The partition size for images with indices greater than the remainder is one less. A global index $1 \le k \le n$ maps to partition index α by the formula,

$$\alpha = \begin{cases} \lfloor \frac{k-1}{m} \rfloor + 1 & r = 0 \\ \lfloor \frac{k-1}{m} \rfloor + 1 & k \le mr \\ \lfloor \frac{k-r-1}{m-1} \rfloor + 1 & k > mr \end{cases} \, . \tag{3.20}$$

The global base index for partition α has the value,

$$k_0^{\alpha} = \begin{cases} (\alpha - 1)m^{\alpha} & r = 0 \quad \alpha = 1, \dots, p \\ (\alpha - 1)m^{\alpha} & r > 0 \quad \alpha = 1, \dots, r \\ (\alpha - 1)m^{\alpha} + r & r > 0 \quad \alpha = r+1, \dots, p \end{cases} \, , \tag{3.21}$$

and the local index has the value,

$$i = k - k_0^{\alpha} \, . \tag{3.22}$$

The local index maps back to the global index according to the rule,

$$k = k_0^{\alpha} + i \, , \quad i = 1, \dots, m^{\alpha} \, . \tag{3.23}$$

An alternative convention is based on the floor function,

$$m = \lfloor n/p \rfloor . \tag{3.24}$$

The partition sizes are the same for all but the last partition according to the formula,

$$m^\alpha = \begin{cases} m & \alpha < p \\ n - (p-1)m & \alpha = p \end{cases} . \tag{3.25}$$

With this alternative convention, the formulas for mapping indices have the values,

$$\alpha = \lfloor \frac{k-1}{m} \rfloor + 1 \tag{3.26}$$

$$k_0^\alpha = (\alpha - 1)m \tag{3.27}$$

$$i = k - k_0^\alpha \tag{3.28}$$

$$k = k_0^\alpha + i \tag{3.29}$$

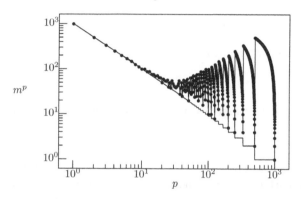

FIGURE 3.1: Partition size based on the floor function as a function of the number of images for a fixed index set $n = 1000$. The solid line is a step function representing the partition size assigned to images not equal to the last image. The bullets represent the partition size assigned to the last image.

For the case $n = 41$, $p = 9$, the original definition of the partition operator yields sizes $(5, 5, 5, 5, 5, 4, 4, 4, 4)$. The alternative definition yields sizes $(4, 4, 4, 4, 4, 4, 4, 4, 9)$. The disadvantage of the alternative definition is that all images must allocate co-array variables with size 9 to accommodate the last image even though all the other images need only size 4. In addition, since the last image owns more data than the others, a workload imbalance may develop as images wait for the last image. This problem becomes more important as the number of images increases as shown in Figure 3.1.

It might be tempting to use the ceiling function in place of the floor function to define the partition size for all but the last image. Formulas for the partition size, however, become more complicated. Cases exist where the partition size is zero for the last image and something smaller than the ceiling function for some images below the last one. This rule can still be used with caution, but it may lead to errors and it may waste one image with nothing to do.

3.3   Row-partitioned matrix-vector multiplication

Matrix-vector multiplication,

$$y = Ax ,\tag{3.30}$$

is an important operation that occurs frequently in parallel application codes. It provides a good example of how partition operators describe parallel algorithms. Application of a partition operator on both sides of Equation (3.30),

$$P^\alpha y = P^\alpha Ax , \quad \alpha = 1,\ldots,p ,\tag{3.31}$$

yields the partitioned equation,

$$y^\alpha = A^\alpha x ,\tag{3.32}$$

for the result vector y^α assigned to image α where the partitioned matrix,

$$A^\alpha = P^\alpha A ,\tag{3.33}$$

consists of the subset of rows assigned to image α. Figure 3.2 shows the matrix-vector multiplication operation partitioned by rows with a one-to-one correspondence between the partition index α and the image index.

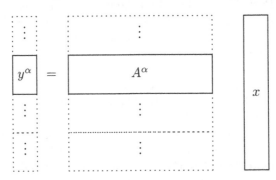

FIGURE 3.2: Row-partitioned matrix-vector multiplication. Image α uses the rows of the partitioned matrix A^α asssigned to it ignoring the other rows of the matrix. It generates a result vector in its local memory.

Listing 3.1 shows code for row-partitioned matrix-vector multiplication.

Once an image has initialized its piece of the matrix and the vector on the right side, it computes its matrix-vector multiplication using only local data. Each image executes the code independently for its own piece of the problem with no interaction between images.

Listing 3.1: Row-partitioned matrix-vector multiplication.

```
integer,parameter  ::  n=globalSize
integer            ::  m
real,allocatable   ::  x(:),y(:),A(:,:)
m = localSize(n)
allocate(A(m,n))
allocate(x(n))
        :           !-code to initialize A and x-!
y = matmul(A,x) !-automatic allocation of result vector-!
```

Listing 3.2 shows the function that computes the size of the partition assigned to the image that invokes it. It is a direct transcription of formula (3.19). Placing this calculation for such a straightforward formula may seem superfluous. But this calculation occurs frequently in a code and is subject to errors that may be hard to find. Furthermore, the programmer can change this function to use an alternative definition for the partition size without changing the rest of the code. In later applications, this kind of function becomes a procedure associated with a class of objects.

Listing 3.2: Local partition size.

```
integer function localSize(n) result(m)
   integer,intent(in)  ::  n
   integer             ::  me,p.r
      p  = num_images()
      me = this_image()
      r  = mod(n,p)
      m  = (n-1)/p + 1
      if(r/=0 .and. me>r) m = m-1
end function localSize
```

The statement,

```
     m = (n-1)/p + 1    !--ceiling(n/p)--!
```

computes the ceiling function for the ratio of two integers.

Partitioned matrices are the basis for optimized versions of serial algorithms with the partition size set equal to the size of a vector register, for example, or to the size of a local cache. These algorithms typically use arrays with one extra dimension corresponding to the partition index with a loop through partitions such that a single processor computes each partition in turn. Indeed, one of the motivations for the co-array model is that the SPMD execution model removes the need for the extra dimension. Each image works on its own piece of the problem, and co-array syntax appears only in specific places to initiate communication between images. The serial version of the code often becomes the basis for the parallel version with the full problem size replaced by the partitioned problem size.

3.4 Input/output in the co-array model

The programmer is still faced with the problem of initializing the data. If the full matrix is small enough to fit in one local memory, the programmer might initialize the matrix by letting one image read the whole matrix into its memory and then let the other images pull their parts into their own local memory. It is more likely, however, that the matrix is too big to fit into a single local memory. A common reason for writing parallel code, however, is that the data objects have become too large for the available local memory.

Listing 3.3: Reading a dense matrix from a file.

```
function readA(filename,n) result(A)
  character(len=*),intent(in)  :: fileName
  integer,intent(in)           :: n
  real,allocatable             :: A(:,:)
  integer                      :: k,k0,m,me,p,r
  real,allocatable :: temp(:)[:]   !..co-array buffer..!
    p  = num_images()
    me = this_image()
    r = mod(n,p)              !..remainder..!
    m = localSize(n)
    k0 = (me-1)*m             !..global base..!
    if (r/=0 .and. me>r) k0=k0+r
    allocate(A(m,n))
    if(me==1) open(unit=10,file='fileName')
    allocate(temp(n)[*]) !..hidden barrier..!
    do k=1,n
       if(me == 1) read(unit=10,*) temp(1:n)
       sync all
       A(1:m,k) = temp(k0+1:k0+m)[1]
       sync all
    end do
    if(me == 1) close(unit=10)
    deallocate(temp)       !..hidden barrier..!
  end function readA
```

Listing 3.3 shows code that uses a temporary co-array buffer to hold each column of the matrix as image one reads one column at a time from a file. Each image executes the **sync all** statement waiting for image one to finish reading the data, and then each image pulls its piece of the matrix into its own local memory. The second `sync all` statement guarantees that image one does not read new data into the buffer before all images have obtained their data. Notice the calculation of the global base index defined by formula (3.21) as each image determines its piece of the current column.

The function shown in Listing 3.3 assigns image one to open a file, read the data, and make data available to the other images. This technique is very common in parallel applications, but serialized input may become a performance bottleneck. To avoid this problem, the programmer may want to associate a procedure pointer to a library procedure that performs input in parallel perhaps from a library using another programming model.

The current version of the co-array model allows only one image at a time to open a particular file. It need not be image one, but only one at a time. Some future version of the language may allow the programmer to use direct access files with each row, or column, of the matrix held in a separate record. Each image could then position itself at the records corresponding to its partition and read from the file independently in parallel. Section A.13 contains a more detailed description of input/output for the co-array model.

3.5 Exercises

1. For the case $n = 1000$ with $p = 37$, use the ceiling function $\lceil n/p \rceil$ to define a base partition size. Use that size for as many images as possible. What is the size on the last image? On the second to last image?

2. Modify the code sample in Listing 3.3 so that image one sends the data to the other images.

3. Modify the code in Listing 3.3 such that image p reads the data from the file.

4. Each image obtains only its own piece of the result vector from the row-partitioned matrix-vector multiplication code. How could each image obtain the full result?

Chapter 4

Reverse Partition Operators

In addition to partitioning data structures in the forward direction, from the global index to the local index, the programmer needs to recombine data structures in the reverse direction, from the local index back to the global index. The partition operators defined in Chapter 3 provided a recipe in the forward direction. Reverse partition operators defined in this chapter provide a recipe in the reverse direction. Almost all operations on data structures found in parallel applications are based on some mixture of forward and reverse partition operations.

4.1 The partition of unity

The set of forward partition operators defined in Chapter 3,

$$P^\alpha , \quad \alpha = 1, \ldots, p , \tag{4.1}$$

labeled with superscript indices, induces a corresponding set of reverse operators,

$$P_\alpha , \quad \alpha = 1, \ldots, p , \tag{4.2}$$

labeled with subscript indices, under the constraint,

$$\sum_{\alpha=1}^{p} P_\alpha P^\alpha = I , \tag{4.3}$$

where I is the $n \times n$ identity operator and n is the size of the global index set.

This constraint is called a partition of unity. It describes the recovery of a global data structure from the local data structures. Indeed, application of the identity operator (4.3) to a vector,

$$\left(\sum_{\alpha=1}^{p} P_\alpha P^\alpha \right) x = x \ , \tag{4.4}$$

changes nothing. Define the local vector,

$$x^\alpha = P^\alpha x \ , \tag{4.5}$$

as before and use the associative property of multiplication to obtain the result,

$$\sum_{\alpha=1}^{p} P_\alpha x^\alpha = x \ . \tag{4.6}$$

To recover the global vector, therefore, each image must apply the reverse operator to its local piece of the vector and then it must sum together pieces from the other images.

The partition of unity is a fundamental tool for the development of parallel algorithms. Whenever a formula, like formula (4.6), involves partition indices not equal to the local partition index, the formula implies communication between images. The mathematical expression for a partitioned algorithm, therefore, explicitly contains the interactions between processors within itself.

The definitions for these operators are not unique, but given the forward operator as in Section 3.1,

$$P^\alpha = \left[\, 0 \quad \cdots \quad I^\alpha \quad \cdots \quad 0 \, \right] \ , \tag{4.7}$$

the reverse operator must be the transposed operator,

$$P_\alpha = \begin{bmatrix} 0 \\ \vdots \\ I_\alpha \\ \vdots \\ 0 \end{bmatrix} \ , \tag{4.8}$$

under the constraint imposed by the partition of unity. This matrix representation of the reverse operator has n rows and m^α columns. The symbol 0 represents a zero matrix and the matrix I_α is the identity matrix for partition α. Applied to a vector x^α of length m^α, the reverse operator produces a vector of length n,

$$\begin{bmatrix} 0 \\ \vdots \\ x^\alpha \\ \vdots \\ 0 \end{bmatrix} = \begin{bmatrix} 0 \\ \vdots \\ I_\alpha \\ \vdots \\ 0 \end{bmatrix} \left[\, x^\alpha \, \right] \ , \tag{4.9}$$

with all zeros except in the part corresponding to partition α. The reader may readily verify, by direct matrix multiplication and summation, that the partition of unity (4.3) is satisfied for these definitions of the forward and reverse operators.

Partition operators are not projection operators. Nor are they inverses of each other. The forward operator maps a vector of length n to a vector of length m^α,

$$P^\alpha : V_n \to V_{m^\alpha} \ . \tag{4.10}$$

The reverse operator maps a vector of length m^α to a vector of length n,

$$P_\alpha : V_{m^\alpha} \to V_n \ . \tag{4.11}$$

The important property of these operators is that the product operators,

$$P_\alpha P^\alpha : V_n \to V_n \ , \tag{4.12}$$

form a set of orthogonal projection operators [79]. Summed over the index α, these projection operators yield the identity operator. Nothing is lost during the forward and reverse partition operations.

The forward partition operation is an example of the scatter operation, and the reverse partition operation is an example of the gather operation. Chapter 5 discusses these operations in more detail implemented as collective operations.

4.2   Column-partitioned matrix-vector multiplication

Many parallel algorithms follow directly from the constraint imposed by the partition of unity. Its insertion between a matrix and a vector

$$y = A \left(\sum_{\alpha=1}^{p} P_\alpha P^\alpha \right) x \tag{4.13}$$

yields a formula for a column-partitioned algorithm for matrix-vector multiplication. The associative rule for multiplication yields the formula,

$$y = \sum_{\alpha=1}^{p} A_\alpha x^\alpha \ , \tag{4.14}$$

where the matrix,

$$A_\alpha = A P_\alpha \ , \tag{4.15}$$

holds the columns assigned to image α, and the vector,

$$x^\alpha = P^\alpha x \ , \tag{4.16}$$

holds the piece assigned to image α. To verify the meaning of the symbol AP_α, observe that,

$$P_\alpha = (P^\alpha)^T , \tag{4.17}$$

implies the identity,

$$AP_\alpha = (P^\alpha A^T)^T . \tag{4.18}$$

The matrix $P^\alpha A^T$ on the right consists of the rows of A^T assigned to the partition α. These rows of the transposed matrix are just the columns of the original matrix A assigned to image α.

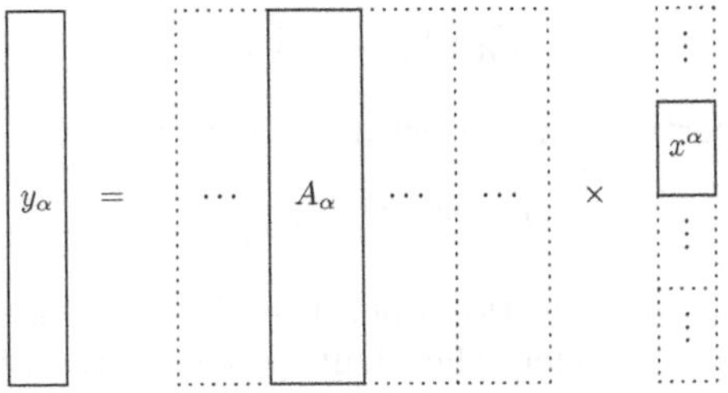

FIGURE 4.1: Column-partitioned matrix-vector multiplication. Image α applies the column-partitioned matrix A_α to the partitioned vector x^α to produce a partial result. The images work independently, but to obtain the full result, each image must sum the partial results from the other images.

Figure 4.1 shows how each image computes a partial result vector,

$$y_\alpha = A_\alpha x^\alpha , \tag{4.19}$$

with length equal to the full dimension. The images do not interact, but to obtain the full result, they must sum together the partial results. Listing 4.1 shows code for the summation. The code employs staggered references to remote images to reduce pressure on local memory should every image try to access the same memory at the same time.

Listing 4.1: Summing the full result for column-partitioned matrix-vector multiplication.

```
integer          :: me,n,p,q
real,allocatable :: buffer(:)[:]
real,allocatable :: y(:)

   :

p  = num_images()
me = this_image()
n  = ...

   :

allocate(y(n))
allocate(buffer(n)[*])

   :
```

```
buffer = matmul(A,x)
y(1:n) = buffer(1:n)
sync all
alpha = me
do q=1,p-1
  alpha = alpha+1
  if(alpha>p) alpha = alpha-p
  y(1:n) = y(1:n) + buffer(1:n)[alpha]
end do
```

4.3   The dot-product operation

Another important application of the partition of unity is calculation of the dot product,

$$d = u^T v \ , \tag{4.20}$$

between two vectors u and v of size n. To obtain a parallel version of this operation, insert the partition of unity (4.3) between the two vectors to obtain

$$d = u^T \left(\sum_{\alpha=1}^{p} P_\alpha P^\alpha \right) v \ , \tag{4.21}$$

and use the associative property of multiplication to write the dot product in partitioned form,

$$d = \sum_{\alpha=1}^{p} (u^T P_\alpha)(P^\alpha v) \ . \tag{4.22}$$

From definition (4.17) for the reverse partition operator, the dot product becomes

$$d = \sum_{\alpha=1}^{p} (P^\alpha u)^T (P^\alpha v) \ . \tag{4.23}$$

Define the partitioned vectors, $u^\alpha = P^\alpha u$ and $v^\alpha = P^\alpha v$, and write

$$d = \sum_{\alpha=1}^{p} (u^\alpha)^T (v^\alpha) \ . \tag{4.24}$$

This formula says that each image first computes its local dot product,

$$d^\alpha = (u^\alpha)^T v^\alpha \ , \tag{4.25}$$

and then participates in a sum,

$$d = \sum_{\alpha=1}^{p} d^\alpha \ , \tag{4.26}$$

across all images such that they all obtain the same value of the dot product.

4.4   Extended definition of partition operators

The previous analysis of parallel algorithms depended only on the fact that the forward and reverse partition operators define a partition of unity. The analysis does not change with a change in the definition of the operators as long as the new operators also define a partition of unity. The formulas for mapping global indices to local indices and local indices to global indices are more complicated but still straightforward.

One way to extend the definition of a partition operator is to create an equivalence class of operators based on a set of orthogonal operators,

$$Q^T Q = Q Q^T = I \ . \tag{4.27}$$

Since the original operators define a partition of unity, this orthogonality condition is unchanged by the insertion of the identity,

$$Q^T \left(\sum_{\alpha=1}^{p} P_\alpha P^\alpha \right) Q = I \ . \tag{4.28}$$

The extended forward operator,

$$Q^\alpha = P^\alpha Q \ , \tag{4.29}$$

and the corresponding reverse operator,

$$Q_\alpha = Q^T P_\alpha = (P^\alpha Q)^T \ , \tag{4.30}$$

still form a partition of unity,

$$\sum_{\alpha=1}^{p} Q_\alpha Q^\alpha = I \ . \tag{4.31}$$

The extended operators mix the elements of a vector before cutting it into pieces,

$$v^\alpha = Q^\alpha v = P^\alpha Q v \ . \tag{4.32}$$

The product operators,

$$Q_\alpha Q^\alpha : V_n \to V_n \ , \tag{4.33}$$

still define a set of orthogonal projection operators. Summing them together unmixes the elements of the partitioned vectors and reassembles the original vector. The analysis of partitioned algorithms does not change; only the specific implementation changes because the index maps are more complicated.

The full generality of these extended operators is seldom used in parallel codes. One particular case, however, where the unitary operator is a permutation,

$$Q = \Pi \ , \tag{4.34}$$

is quite common. It occurs in Chapter 13 for cyclic distributions and again in Chapter 14 for finite element meshes.

4.5 Exercises

1. Verify that the forward operators (4.7) and reverse operators (4.8) satisfy the partition of unity (4.3).

2. Given the definition of the forward partition operators, show that there is one and only one definition for the reverse partition operators that satisfies the partition of unity.

3. Show that the set of operators,

$$Q_\alpha Q^\alpha \quad \alpha = 1, \ldots, p ,$$

defined by (4.33), is a set of orthogonal projection operators.

Chapter 5

Collective Operations

Collective operations calculate such quantities as vector norms and error residuals or the value of some physical quantity such as the value of the energy to check conservation of energy. Because these operations involve communication across images, they must be implemented carefully and used sparingly. Naive implementations based on a linear algorithm limit scalability because execution time increases linearly with the number of images. Implementations based on a binary-tree algorithm improve scalability because the execution time increases more slowly like the logarithm of the number of images. Optimization of collective operations is an active area of research [11] [18] [66] [96]. Indeed, for very large machines, it may be necessary to design new algorithms free of collective operations [4].

5.1 Reduction to root

The reduction operation is a fundamental collective operation. At the beginning, each image holds its own value of a variable. At the end, the root image holds the reduced value, commonly the sum, the maximum or the minimum value across images. Listing 5.1 shows a sum-to-root operation using a linear algorithm involving just the root image. Although the execution time increases linearly with processor count, it may be acceptable if it does not occur in a time-critical part of the application. If it does occur in a time-critical part, however, or if it is used frequently, it may limit performance in unacceptable ways.

Listing 5.1: Code for linear summation to the root image.

```
integer :: i,me,p,root
real    :: s
real    :: x[*]
p     = num_images()
me    = this_image()
      :
x     = someValue
sync all
if(me == root) then
   s = x
   do i = 2,p
      s = s + x[i]
   end do
end if
```

Figure 5.1 shows the logical flow of data up a binary tree with image one acting as the root image. The tree has $k = \lceil \log_2 p \rceil$ stages where $\lceil \cdot \rceil$ is the ceiling function. With $p = 13$, there are four stages. At stage one, images $(1, 3, 5, 9, 11)$ access values from partners $(2, 4, 6, 10, 12)$ with image index one larger than their own. Since the number of images is odd, the last image has no partner and does not participate.

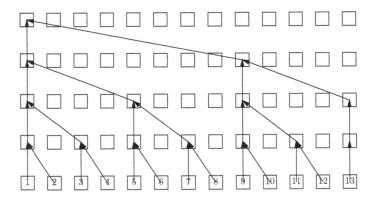

FIGURE 5.1: Data flow for a reduction operation up a binary tree for the case $p = 13$ images.

At stage two, images $(1, 5, 9)$ pair with partners $(3, 7, 12)$ with image index two larger. Again the last image does not participate. At stage three, images $(1, 9)$ pair with partners $(5, 13)$ with index four larger. The last image finally participates at this stage. At stage four, the root image (1) pairs with image (9) eight larger. The root image participates at all stages.

Listing 5.2 shows code for the sum-to-root function using a binary-tree algorithm. The incoming dummy argument x has the `intent(in)` attribute so it cannot be changed. The reduced value is the sum of the input values across all images, and this value is returned in the result variable s on the root image. If the optional argument root is not present, image one is the default root image.

Listing 5.2: Code for the sum-to-root function using a binary-tree algorithm.

```
real function sumToRoot(x,root) result(s)
implicit none
real,intent(in)                 :: x
integer,optional,intent(in)     :: root
real,allocatable                :: y[:]
integer                         :: L,me,p
   p  = num_images()
   me = this_image()
   allocate(y[*])
   y  = x
   sync all
   L = 1
   do while(L < p)
     if(me+L<=p .and. mod(me-1,2*L)==0) y=y+y[me+L]
     L = 2*L
     sync all
   end do
   s = 0.0
   if(present(root)) then
      if(me == root) s = y[1]
   else
      if(me == 1)    s = y
   end if
end function sumToRoot
```

To make the value of the incoming variable visible across images, each image copies its value to the co-array variable y. After executing the sync all statement to ensure that all images have made their value visible, the function enters a loop containing logic for traversing the tree. The logical value me+L<=p ensures that partners at each stage are not beyond the number of images, and the logical value mod(me-1,2*L) ensures that only those images with the correct indices participate at each stage. The sync all statement inside the loop ensures that each stage is complete before going to the next stage.

At the end of the loop, image one holds the sum. If it is the default root image, it places that value in the return variable. If the optional second argument is present, the image with index equal to the root obtains the reduced value from image one and places it in the return variable. Only the root image receives a valid return value. The other images return the value zero. Alternatively, the function could return a NaN value to avoid inadvertent use of the result by images not equal to the root.

Adding floating-point numbers is subject to rounding errors that depend on the order of accumulation. For the linear algorithm shown in Listing 5.1, for example, the result may be different if the loop traverses image indices in reverse order. Likewise, the binary-tree algorithm shown in Listing 5.2 accumulates the result in an order that depends on the binary tree and may be different from the value obtained by the linear algorithm. If an application is highly sensitive to rounding errors, the programmer might want to consider other algorithms to decrease the sensitivity.

5.2   Broadcast from root

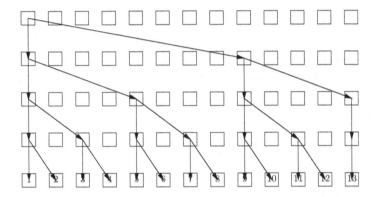

FIGURE 5.2: Data flow for the broadcast operation down a binary tree for the case $p = 13$ images.

The broadcast operation is another fundamental collective operation. At the beginning, the root image holds the value of a variable. At the end, all the images hold the value. Figure 5.2 shows the logical flow of data from the root down the binary tree, the same logic as shown in Figure 5.1 with the arrows reversed. Listing 5.3 shows the code from Listing 5.2 converted into the broadcast function.

Listing 5.3: The root-to-all function.

```
real function rootToAll(x,root) result(s)
implicit none
real,intent(in)                    :: x
integer,optional,intent(in)    :: root
real,sllocable                     :: y[:]
integer                            :: L,me,p
  p  = num_images()
  me = this_image()
  allocate(y[*])
```

```
if(present(root)) then
   if(me == root) y[1] = x
else
   if(me == 1) y = x
end if
sync all
L = 1
do while(L < p)
   L = 2*L
end do
do while(L > 0)
   if( me+L<=p .and. mod(me-1,2*L)==0) y[me+L]=y
   L = L/2
   sync all
end do
s = y
end function rootToAll
```

5.3 The sum-to-all operation

Composition of the sum-to-root function with the root-to-all function yields a function that returns the global sum to every image as shown in Listing 5.4. One reason for writing collective operations as functions is that they can be composed in this way to obtain other functions. The sum-to-all operation is one of the most important operations found in parallel application codes. Chapter 6 discusses how to develop a model for the execution time for this operation.

Listing 5.4: The sum-to-all function.

```
real function sumToAll(x) result(s)
   real,intent(in) :: x
   s = rootToAll(sumToRoot(x))
end function sumToAll
```

Listing 5.5 shows how to use the sum-to-all function to perform the dot-product operation as described previously in Chapter 4. Each image computes its local dot product,

$$d^\alpha = (u^\alpha)^T \cdot v^\alpha \,, \tag{5.1}$$

followed by a global-sum operation,

$$d = \sum_{\alpha=1}^{p} d^\alpha \,. \tag{5.2}$$

The result is the composition of two functions, the intrinsic Fortran function and the sum-to-all function.

Listing 5.5: The dot-product operation.

```
d = sumToAll(dot_product(u,v))
```

5.4 The max-to-all and min-to-all operations

In a similar way, the binary-tree algorithm is useful for finding the maximum value of a vector,

$$s = \max_{k=1,n} x_k \; . \tag{5.3}$$

Since the result is independent of the order that the elements are examined, the maximum value over two indices,

$$s = \max_{\alpha=1,p} \left(\max_{i=1,m^\alpha} x_i^\alpha \right) , \tag{5.4}$$

one for the partition index and one for the local index, is the same as the maximum value over the single global index. Each image first computes the maximum value for its local vector and then participates in a collective operation to obtain the maximum across images. The same analysis holds for finding the minimum by replacing the maximum with the minimum in the above equations. Code for the max-to-root function is a straightforward modification of the code for the sum-to-root function.

Listing 5.6: The max-to-all function.

```
real function maxToAll(x) result(s)
   real,intent(in) :: x
   s = rootToAll(maxToRoot(x))
end function maxToAll
```

Listing 5.6 shows the composition of this function with the root-to-all function to obtain the max-to-all function. Since $\min_k(x_k) = -\max_k(-x_k)$, the min-to-all function is just an invocation of the max-to-all function with appropriate sign changes as shown in Listing 5.7.

Listing 5.7: The min-to-all function.

```
real function minToAll(x) result(s)
   real,intent(in) :: x
   s = -maxToAll(-x)
end function minToAll
```

5.5 Vector norms

Listing 5.8 shows how to compute vector norms using collective functions.

Listing 5.8: Vector norms.

```
real  :: L_1,L_2,L_inf
real,allocatable :: u(:)
      :
  L_1   = sumToAll(sum(abs(u)))
  L_2   = sqrt(sumToAll(dot_product(u,u)))
  L_inf = maxToAll(maxval(abs(u)))
```

5.6 Collectives with array arguments

It is straightforward to modify these binary-tree algorithms to accept as input an array variable and to return as output an array variable containing the pointwise reduced values. Interface blocks, like the ones shown in Listing 5.9, resolve the generic name for different versions.

Listing 5.9: Interface block for the sum-to-all function.

```
interface sumToAll
  function sumToAll0(x) result(s)
    real,intent(in) :: x
    real            :: s
  end function sumToAll0

  function sumToAll1(x) result(s)
    real,intent(in)   :: x(:)
    real,allocatable :: s(:)
  end function sumToAll1
end interface sumToAll
```

The sum-to-all function with an array argument can accumulate the results of a column-partitioned matrix-vector multiplication as shown in Listing 5.10, replacing the code shown in Listing 4.1. The collective function provides all the details and the required synchronization.

Listing 5.10: Summing the full result for column-partitioned matrix-vector multiplication using the sum-to-all function.

```
y = sumToAll(matmul(A,x))
```

Similarly, the root-to-all function can broadcast an array read from a file to the other images as shown in Listing 5.11.

Listing 5.11: Broadcasting an array.

```
if(me == 1) read(*,*) x(:)
x = rootToAll(x)
```

5.7 The scatter and gather operations

The scatter and gather operations are important examples of collective operations [42, p. 103, p. 164]. The scatter operation is equivalent to the forward partition operation, and the gather operation is equivalent to the reverse partition operation.

Listing 5.12 shows code for the scatter operation. At the beginning, the root image holds an array of global values and at the end of the operation, each image holds a partitioned array with values corresponding to its image index. The root image copies the input array into a co-array variable to make it visible across images. Then each image determines the values it owns and reads them into a local array variable. Alternatively, the root image could distribute partitioned arrays across images.

Listing 5.12: The scatter operation.

```
module ClassScatter
  integer,parameter,private :: wp=kind(1.0d0)
  interface scatter
    procedure scatterFromRoot
  end interface scatter
contains
 function scatterFromRoot(x,rute) result(y)
   implicit none
   real(wp),intent(in)             :: x(:)
   integer,intent(in),optional :: rute
   real(wp),allocatable        :: y(:)
   real(wp),allocatable        :: temp(:)[:]
   integer,allocatable         :: n[:]
   integer :: me,p
   integer :: k0,m,r,root
   p  = num_images()
   me = this_image()
   allocate(n[*])
   if(present(rute)) then
     root = rute
   else
```

```
      root = 1
   end if
   if(me==root) n = size(x)
   sync all
   n = n[root]
   allocate(temp(n)[*])
   if(me == root) temp = x
   r   = mod(n,p)
   m   = (n-1)/p+1            !....ceiling....!
   k0 = (me-1)*m
   if(r/=0 .and. me>r) then
      m   = m-1
      k0 = (me-1)*m+r
   end if
   sync all
   y = temp(k0+1:k0+m)[root]
   deallocate(temp)
  end function scatterFromRoot
end module ClassScatter
```

Listing 5.13 shows code for the gather operation. At the beginning, each image holds values in a local array variable and, at the end, the root image holds an array with the values from each image in order.

Listing 5.13: The gather operation.

```
module ClassGather
  integer,parameter,private :: wp=kind(1.0d0)
  interface gather
    procedure gatherToRoot
  end interface gather
contains
 function gatherToRoot(x,rute) result(y)
   use collectives
   implicit none
   real(wp),intent(in)            :: x(:)
   integer,intent(in),optional :: rute
   real(wp),allocatable           :: y(:)
   real(wp),allocatable           :: temp(:)[:]
   integer :: me,p
   integer :: k0,m,n,r,root
   p   = num_images()
   me = this_image()
   m   = size(x)
   n   = sumToAll(m)
   if(present(rute)) then
     root = rute
   else
     root = 1
   end if
```

```
   allocate(temp(n)[*])
   r  = mod(n,p)
   k0 = (me-1)*m
   if(r/=0 .and. me>r) then
      k0 = (me-1)*m+r
   end if
   temp(k0+1:k0+m)[root] = x(1:m)
   sync all
   allocate(y(n))
   if(me == root) then
      y = temp
   else
      y = 0.0_wp
   end if
   deallocate(temp)
 end function gatherToRoot
end module ClassGather
```

Composition of the gather-to-root function with the root-to-all function yields the gather-to-all function shown in Listing 5.14. The input array from each image is the local array assigned to the image, and the output array holds the global array of concatenated local arrays ordered by the image indices.

Listing 5.14: The gather-to-all function.

```
function gatherToAll(x) result(s)
   real,allocatable :: s(:)
   real,intent(in)  :: x(:)
   s = rootToAll(gatherToRoot(x))
end function gatherToAll
```

Another collective operation is called the all-to-all operation that sometimes occurs in implementations of the transpose operation [11] [42, p. 164]. Listing 5.15, for example, shows code that transposes a square matrix of size p, partitioned by rows, to the transposed matrix, also partitioned by rows, using the scalar version of the gather-to-root function. Chapter 10 discusses the transpose operation in more detail.

Listing 5.15: The transpose operation using the gather operation.

```
integer :: me,p,q
real,allocatable :: temp(:),x(:)
p=num_images()
me=this_image()
allocate (temp(p))
allocate (x(p))
   :
do q=1,p
  temp = gatherToRoot(x(q),q)
  if(me == q) y = temp
```

```
end  do
```

5.8   A cautionary note about functions with side effects

The collective functions have side effects because internal synchronization is not visible to the calling procedure. Using these functions in statements that combine results from two or more functions may lead to unexpected behavior. The Fortran language allows the compiler to evaluate functions within an expression in any order that would give the same result in exact arithmetic or even to skip invocation of a function if it can determine its value without invoking it. If some images skip invocation of a function while others invoke it, the code may deadlock. If different images invoke the functions within an expression in different order, one image might invoke the synchronization statement in one function while another image invokes the synchronization statement in another function. Such behavior may create a livelock condition where the program advances but produces indeterminant results. Errors like these are very difficult to diagnose.

The strength of implementing collective operations as functions is that they can be composed to form other functions. In addition, two functions can be written as inverses of each other. The gather operation, for example, is the inverse of the scatter function where execution of the statement shown in Listing 5.16 leaves the array variable x unchanged. Problems from side effects do not occur for these composed functions.

Listing 5.16: The gather function is the inverse of the scatter function.

```
x = gather(scatter(x))
```

Deciding how to implement collective operations is a matter of personal preference and programming style. The programmer can eliminate problems created by side effects by replacing functions with subroutines.

5.9   Exercises

1. Draw the picture, shown in Figure 5.1 for the binary-tree algorithm, for the case $p = 23$.

2. In the sum-to-root function shown in Listing 5.2, why is it safe for an image to accumulate the new value y directly into the same variable?

3. Modify the sum-to-root function shown in Listing 5.2 to obtain the max-to-root function.

4. Combine the max-to-root function with the root-to-all function to obtain the max-to-all function.

5. Modify the sum-to-root function shown in Listing 5.2 to accept an array variable as input and to return an array variable holding the pointwise sum.

6. Implement a sum-to-all function that accepts a two-dimensional array as input and returns the pointwise sum of the elements of the array as a two-dimensional array. How must the interface block be altered?

7. Modify the code for the gather-to-root function so it uses a binary-tree algorithm. Which version would you expect to perform better?

8. Verify that the scatter and gather functions are inverses of each other.

9. How are the scatter operation and the forward partition operation related to each other? How are the gather operation and the reverse operation related?

10. Verify that the code shown in Listing 5.10 produces the full result from the row-partitioned matrix-vector multiplication shown in Listing 3.1. Fill in the details required to make the code work.

Chapter 6

Performance Modeling

A computer performance model is a mathematical formula describing execution time for a specific algorithm usually as a function of problem size and processor count. The formula typically involves several hardware and software parameters, and it is subject to change as computer architectures change. A model that works well for a current machine may not work well for a new version of the machine even from the same vendor. A model that contains the main features that influence execution time, incomplete as it may be, informs a realistic expectation for the performance of a specific algorithm on a specific machine.

Most performance models are modified versions of Amdahl's Law. Execution time decreases as the processor count increases, but the decrease is limited by parts of an algorithm where the execution time does not decrease. For a fixed problem size, called the strong-scaling case, this behavior leads to a pessimistic view of parallel processing. In practice, however, as the processor count increases, programmers often solve bigger problems. This weak-scaling case corresponds to three different constraints: the fixed-time constraint, the fixed-work constraint, and the fixed-efficiency constraint.

This book reports all measured execution times as normalized values relative to some reference time. Absolute execution times cannot be recovered from these values and cannot be tied to a specific machine. Comparing different machines is a delicate topic called benchmarking and is beyond the scope of this book [52].

6.1 Execution time for the sum-to-all operation

The sum-to-all operation is an apparently simple operation that nonetheless reveals many of the principles involved in computer performance modeling. Figure 6.1 shows measured execution times for the sum-to-all function shown in Listing 5.4. The bullets mark average values for 1000 invocations of the function normalized to the time on a single image. The measured times show a kink near image count $p_0 = 24$ because the machine has a node structure with 24 physical processors in each node.

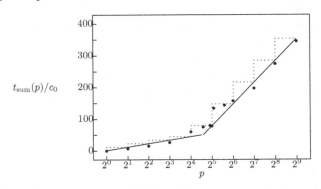

FIGURE 6.1: Execution time normalized to the execution time on one image for the sum-to-all operation as a function of the image count. Each invocation of the sum-to-all function performs a global sum of one double-precision variable, and the bullets (•) represent the average time over 1000 invocations. The dotted lines represent the timing formula (6.1) using the parameters defined in (6.2). The solid line represents formula (6.3) ignoring the step functions.

The sum-to-all function employs a binary-tree algorithm, and the execution time is dominated by synchronization statements at each stage up the tree and each stage down the tree. There are $\lceil \log_2 p \rceil$ stages in each direction, and the synchronization time outside a node is longer than inside a node. These observations, along with the measured execution times, suggest that the function,

$$t_{\text{sum}}(p) = 2\tau_k \lceil \log_2 p \rceil + c_k , \quad k = 0, 1 , \tag{6.1}$$

is a good candidate for a performance model. The two branches of the function correspond to a short synchronization time τ_0 within a node, and a longer synchronization time τ_1 outside a node. The constants that appear in the formula,

$$
\begin{aligned}
c_0 &= t_{\text{sum}}(1) & 1 \le p \le p_0 , \\
c_1 &= c_0 - 2(\tau_1 - \tau_0)\lceil \log_2 p_0 \rceil & p_0 \le p \le p_1 ,
\end{aligned}
\tag{6.2}
$$

enforce continuity across the kink although continuity is not necessarily required.

A model that captures the important features of the execution time, without obscuring them with relatively minor details, is more useful than one that attempts to explain every detail without adding much to the explanation of observed behavior. In particular, removing the step function in the formula for execution time,

$$t_{\text{sum}}(p) = 2\tau_k \log_2 p + c_k , \quad k = 0, 1 , \tag{6.3}$$

deletes little from the analysis, especially if the number of images is a power of two, where the formula is correct as shown by the solid lines in Figure 6.1. The missing measurements between powers of two are easy to estimate by drawing in the step functions like the dotted lines in the figure.

The normalized execution time, relative to the time on a single image, obeys the formula,

$$t_{\text{sum}}(p)/c_0 = \begin{cases} 1 + 2(\tau_0/c_0) \log_2 p & 1 \leq p \leq p_0 \\ 1 + 2(\tau_1/c_0) \log_2 p - 2(\tau_1/c_0 - \tau_0/c_0) \log_2 p_0 & p_0 \leq p \leq p_1 \end{cases} .$$

For this particular machine, the values $\tau_0/c_0 = 5.5$ and $\tau_1/c_0 = 34.2$ provide a reasonable fit to the measured results. Slightly different values might work just as well without providing a significantly different picture of how execution time changes with image count.

6.2 Execution time for the dot-product operation

A performance model for the execution time for the dot-product operation is the sum of two terms,

$$t(n, p) = t_{\text{local}}(n, p) + (1 - \delta_1^p) t_{\text{sum}}(p) . \tag{6.4}$$

The first term is the execution time for a local dot-product calculation, and the second term is the execution time for a sum-to-all operation already represented by formula (6.3). The factor in front of the second term containing a delta function reflects the fact that the sum-to-all operation is not invoked on a single image.

The problem size and the image count enter into the formula as independent variables. The problem size itself may be a function of the image count, and different constraints on how the problem size changes with image count lead to different behavior of the execution time as a function of image count.

The execution time for a local dot-product calculation equals the work done divided by the rate of doing work,

$$t_{\text{local}}(n, p) = \left(\frac{2n}{p} \right) \left(\frac{e_0}{r_0} \right) , \tag{6.5}$$

ignoring cases where the problem size is not a multiple of the image count. Omitting this minor detail does not change the basic conclusions of the analysis.

The unit of work, $e_0 = 1$ flop, is the work required to produce one 64-bit floating-point operation, and the computational power r_0 is the rate of doing work measured in flop/s. The local execution time (6.5) plus the sum-to-all time (6.3) yields the total execution time,

$$t(n,p) = \left(\frac{1}{p}\right)\left(\frac{2ne_0}{r_0}\right) + (1 - \delta_1^p)t_{\text{sum}}(p) \ . \tag{6.6}$$

The execution time on one image has the value,

$$t(n,1) = \frac{2ne_0}{r_0} \ , \tag{6.7}$$

so that relative execution time,

$$\tau(n,p) = t(n,p)/t(n,1) \ , \tag{6.8}$$

obeys the formula,

$$\tau(n,p) = 1/p + (1 - \delta_1^p)/\gamma(n,p) \ , \tag{6.9}$$

where the function,

$$\gamma(n,p) = t(n,1)/t_{\text{sum}}(p) \ , \tag{6.10}$$

is called the granularity function. For the dot-product operation, the granularity function has the value,

$$\gamma(n,p) = t(n,1)/\left[2\tau_k \log_2 p + c_k\right] \ , \quad k = 0,1 \ . \tag{6.11}$$

Granularity functions are important quantities in performance analysis. The larger the granularity function, the better the normalized execution time approaches perfect scaling like $1/p$,

$$\mid \tau(n,p) - 1/p \mid = 1/\gamma(n,p) \ . \tag{6.12}$$

In this case, the granularity function measures the importance of the overhead incurred by the sum-to-all operation relative to the local dot-product operation. The smaller the overhead, the larger the granularity function. For a fixed value of p, the function increases with problem size because the execution time on a single image increases. The normalized execution time, therefore, approaches perfect scaling in the limit of large problem size. On the other hand, for a fixed problem size, the function decreases with image count because the number of stages in the binary tree increases. The distance of the execution time from perfect scaling increases for large image counts.

6.3   Speedup and efficiency

The speedup function is the reciprocal of the normalized execution time,

$$s(n, p) = 1/\tau(n, p) \; . \tag{6.13}$$

Although it holds no new information, it is a popular way to display performance measurements. The efficiency function,

$$E(n, p) = s(n, p)/p \; , \tag{6.14}$$

measures the fraction of the best possible speedup achieved by an algorithm.

Different forms of the speedup and efficiency functions result from different choices for how the problem size changes with image count. One choice corresponds to the strong-scaling case where the problem size is fixed. Other choices correspond to weak-scaling cases where the problem size changes with image count.

6.4   Strong scaling under a fixed-size constraint

Strong scaling corresponds to the constraint that the problem size is fixed as the image count changes. Figure 6.2 shows measured execution times, marked with bullets, relative to the time on one image. The solid lines correspond to the model function (6.9) using the same parameters determined independently for the sum-to-all operation already used in Figure 6.1.

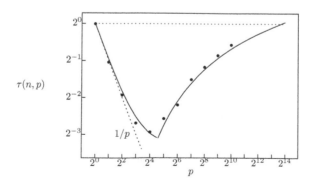

FIGURE 6.2: Normalized execution time as a function of image count for the dot-product operation with fixed problem size $n = 2^{16}$. Bullets (•) mark the measured times. Solid lines represent the model function (6.9).

As the results show, the performance model represents the measured values well. This agreement validates the assumption that the execution time is truly the sum of two independent terms. Had the agreement not been good, the performance model would need to change. For small values of p, the granularity function is large and the execution time decreases like $1/p$. As p increases, the sum-to-all time increases and the granularity function decreases until it is small enough that the execution time starts to increase.

The two-term performance model invites further analysis. Finding the location of the minimum execution time, for example, is an exercise in the calculus with a mild complication arising from the fact that the function has two branches, one inside the node and one outside. For each branch, the minimum occurs when the total derivative with respect to image count equals zero,

$$d\tau/dp = 0 \ . \tag{6.15}$$

Since the problem size is constant, the partial derivative with respect to problem size is zero, and the condition for a minimum yields the formula,

$$p^2 = \frac{\gamma(p)^2}{d\gamma/dp} \ . \tag{6.16}$$

Substitution of the granularity function from (6.11) and the single-image execution time from (6.6) yields the image count at the minimum,

$$p_{\min} = \left(\frac{1}{\log_2 e}\right)\left(\frac{ne_0}{\tau_k r_0}\right) \ , \quad k = 0, 1 \ . \tag{6.17}$$

The minimum value depends on the problem size and on the branch of the function.

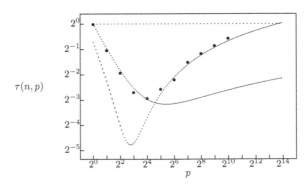

FIGURE 6.3: Two branches of the normalized execution time shown as dotted lines.

Figure 6.3 shows the two branches of the function extended outside their valid ranges. The minimum for the first branch may occur outside the node and the mimimum for the second branch may occur inside the node. In that case, the minimum occurs at the intersection of the two branches at $p_{min} = p_0$ as happens for the particular problem size corresponding to the measurements shown in the figure.

The image count at the minimum is the ratio of the global problem size n to characteristic problem sizes,

$$n_k = (\tau_k r_0/e_0) \log_2 e , \quad k = 0, 1 , \tag{6.18}$$

depending on whether the minimum occurs inside or outside the node. The minimum occurs inside the node for problem sizes in the range,

$$n_0 \leq n \leq n_0 p_0 , \tag{6.19}$$

and outside the node for the range,

$$n \geq n_1 p_0 . \tag{6.20}$$

A larger problem size relative to the characteristic size implies that more images can be used before reaching the minimum execution time.

For large values of p, the execution time increases until eventually it equals the original execution time on a single processor, $t_{sum}(p) = t(n, 1)$, that is when the granularity function equals one,

$$\gamma(n, p) = 1 . \tag{6.21}$$

From definition (6.11), this value occurs when

$$\log_2 p = \frac{t(n, 1) - c_1}{2\tau_1} . \tag{6.22}$$

Beyond this point, the execution time on p images is slower than the execution time on one. Notice that the maximum number of images that can be used for the strong-scaling case equals the problem size, $p_{max} = n$, because the local partition size must be at least one, $n/p \geq 1$.

The left side of Figure 6.4 shows the strong speedup function. This figure is just an upside down version of Figure 6.2 because the two functions are reciprocals of each other. The speedup function for the dot-product operation has the value,

$$s(n, p) = \frac{p}{1 + p(1 - \delta_1^p)/\gamma(n, p)} . \tag{6.23}$$

When the image count reaches the value where the granularity equals one, the speedup is less than one,

$$s(n, p) = \frac{p}{1 + p} , \tag{6.24}$$

after which the speedup becomes a slowdown.

The right side of Figure 6.4 shows the efficiency function,

$$E(n, p) = \frac{1}{1 + p(1 - \delta_1^p)/\gamma(n, p)} . \tag{6.25}$$

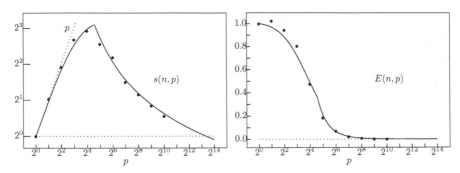

FIGURE 6.4: On the left, the strong speedup function marked by bullets (•). The solid lines represent Equation (6.23) with the same parameters as in Figure 6.2. The diagonal dotted line represents perfect speedup. The right side shows the efficiency function with the solid line representing Equation (6.25).

The model function does not yield very accurate values for the efficiency function for small values of p. In fact, the measured results exhibit super-linear speedup for $p = 2$. This effect typically happens when the partition size decreases enough to fit in the local cache where it had not fit for smaller values of p. The computational power increases because the overhead caused by memory traffic decreases, and the execution time decreases more than would be expected from just adding another image. This result calls into question the assumption of constant computational power, but modeling the behavior of the computational power in minute detail is a somewhat complicated task and yields little insight into the behavior of the execution time.

The behavior of the execution time as a function of image count is an example of a generalized form of Amdahl's Law [1]. The classical statement of the law refers to the case where the execution time is a function of two terms with one term decreasing like the reciprocal of p, while the other term is constant. For the dot-product operation, the first term decreases like $1/p$ but the second term increases like $\log_2 p$.

The disappointing results for the strong-scaling case are typical for algorithms that are not completely parallel, which includes almost all algorithms. For a long time, poor strong-scaling results discouraged some people from developing parallel algorithms. Practitioners of the art, therefore, changed their focus to a new measure of performance called weak scaling.

6.5   Weak scaling under a fixed-time constraint

The disappointing results obtained with strong scaling led to the idea of weak scaling. Rather than being fixed, the problem size changes with image count,

$$n = n(p) \,, \tag{6.26}$$

defined by some function of the image count. One way to specify this function is to impose a fixed-time constraint. It is based on the observation that a programmer may want to solve a bigger problem in about the same time rather than the same problem in less time [44] [45] [46]. More precisely, after measuring the execution time on one image for some arbitrary problem size, $n(1)$, the problem size must change such that the execution time remains the same as the image count changes,

$$t(n(p), p) = t(n(1), 1) \,. \tag{6.27}$$

This constraint provides an implicit formula for the problem size as a function of image count.

To find this function, observe that the computational power for problem size $n(1)$ on one image, under the fixed-time constraint, has the value,

$$r_0 = \frac{2n(1)e_0}{t(n(p), p)} \,. \tag{6.28}$$

On the other hand, the computational power for problem size $n(p)$ on one image has the value,

$$r_0 = \frac{2n(p)e_0}{t(n(p), 1)} \,. \tag{6.29}$$

Under the assumption that the computational power is constant, equating these two values yields the formula,

$$\frac{n(p)}{n(1)} = \frac{t(n(p), 1)}{t(n(p), p)} \,. \tag{6.30}$$

The right side of this formula is the speedup function (6.13), and the relative change in problem size equals the speedup function,

$$n(p)/n(1) = s(n(p), p) \,. \tag{6.31}$$

The difficulty with this formula is that the function $n(p)$ must already be known to apply it.

An explicit formula for the problem size results from the observation that, under the fixed-time constraint, the granularity function obeys the identity,

$$\gamma(n(p), p) = s(n(p), p)\gamma(n(1), p) \,. \tag{6.32}$$

The reciprocal of the speedup function, therefore, obeys the identity,

$$s(n(p), p)^{-1} = 1/p + s(n(p), p)^{-1}(1 - \delta_1^p)/\gamma(n(1), p) \ . \qquad (6.33)$$

Relacement of the speedup function on the right side of Equation (6.31) with the speedup function obtained from (6.33) yields the formula,

$$n(p)/n(1) = p \cdot [1 - (1 - \delta_1^p)/\gamma(n(1), p)] \ . \qquad (6.34)$$

Evaluation of the granularity function that appears in this formula,

$$\gamma(n(1), p) = \frac{t(n(1), 1)}{t_{\text{sum}}(p)} \ , \qquad (6.35)$$

requires just the initial measurement of execution time on one image.

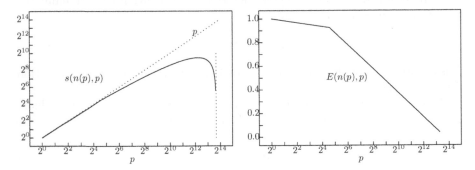

FIGURE 6.5: On the left, the speedup function under the fixed-time constraint. The solid line represents formula (6.34) using the same parameters as in Figure 6.2. The diagonal dotted line represents perfect speedup. The vertical dotted line marks the image count where the partition size equals zero. The right side of the figure shows the corresponding efficiency function.

The left side of Figure 6.5 shows the speedup function under the fixed-time constraint. Equivalently, it shows the change in problem size as the image count changes. For small values of p, the speedup function tracks perfect speedup rather closely because the sum-to-all time within a node is small relative to the local dot-product time and the granularity function is large compared with one. The sum-to-all time, however, increases with the value of p until the granularity function equals one. At that point, both the speedup function and the local partition size,

$$n(p)/p = n(1) \cdot [1 - (1 - \delta_1^p)/\gamma_{n(1)}(p)] \ , \qquad (6.36)$$

equal zero as marked by the vertical dotted line in the figure. This point is the same point where the strong speedup function becomes a slowdown function, as displayed in Equation (6.21).

The right side of Figure 6.5 shows the corresponding efficiency function.

Although the weak efficiency function yields higher efficiency for large p than the strong efficiency function shown in Figure 6.4, the algorithm still achieves only about 10% efficiency at the maximum speedup.

The partition size assigned to each image must decrease with image count to keep the execution time constant because the sum-to-all time increases with the image count. But it decreases less rapidly for the fixed-time case than for the strong-scaling case. Indeed, formula (6.31) yields the result,

$$n(p)/p = s(n(p), p) \cdot (n(1)/p) . \tag{6.37}$$

Since the speedup function is greater than one, the partition size for fixed-time scaling remains larger than the partition size for strong scaling with fixed problem size $n = n(1)$. But it too eventually decreases to zero limiting the number of images that can be used.

6.6 Weak scaling under a fixed-work constraint

Another form of weak scaling employs a fixed-work constraint,

$$2(n(p)/p)e_0 = 2(n(1)/1)e_0 , \tag{6.38}$$

where the local floating-point work remains constant. The local partition size, therefore, also remains constant,

$$n(p)/p = n(1) , \tag{6.39}$$

while the global problem size grows linearly with p. For the dot-product operation, the fixed-work constraint could also be called the fixed-memory constraint. This equivalence is not always true. The programmer claiming to use weak scaling must clearly state which constraint applies.

The assumption of constant computational power implies that the execution time on a single image is a linear function of image count,

$$t(n(p), 1) = pt(n(1), 1) . \tag{6.40}$$

The normalized execution time, therefore, obeys the formula,

$$\tau(n(p), p) = (1/p)[1 + (1 - \delta_1^p)/\gamma(n(1), p)] , \tag{6.41}$$

and the speedup function obeys the formula,

$$s(n(p), p) = p/[1 + (1 - \delta_1^p)/\gamma(n(1), p)] , \tag{6.42}$$

the reciprocal of the normalized execution time.

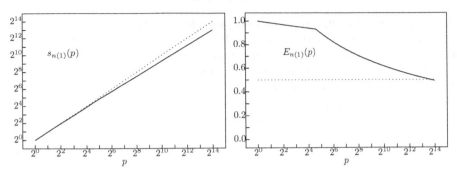

FIGURE 6.6: On the left, the speedup function for the dot-product operation under the fixed-work constraint. The solid line represents formula (6.42) using the same parameters as in Figure 6.5. The diagonal dotted line represents perfect speedup. On the right, the efficiency function. The horizontal dotted line marks the point of half efficiency.

The left side of Figure 6.6 shows the speedup function under the fixed-work constraint. For small values of p, the local execution time is much larger than the sum-to-all time, and the function exhibits near-perfect speedup. Indeed, for small values of p where the granularity function is large, the Taylor expansion of the function yields the formula,

$$s(n(p),p) = p[1 - (1 - \delta_1^p)/\gamma(n(1),p) + O(\gamma(n(1),p)^{-2})] . \qquad (6.43)$$

It differs from the fixed-time speedup function (6.34) only in the second-order term.

Figure 6.6, on the right side, shows the efficiency function. Changing the problem size under the fixed-work constraint uses the machine more efficiently than either strong scaling or fixed-time scaling. The advantage of the fixed-work constraint is that it is defined for all values of p unlike the other two cases. Even for very large values of p, however, the efficiency of the algorithm is still less than half.

6.7   Weak scaling under a fixed-efficiency constraint

Yet another way to pick the problem size for weak scaling is to impose a fixed-efficiency constraint also called an iso-efficiency constraint [67] [81]. From definition (6.14), the speedup function under this constraint is a linear function of the image count,

$$s(n(p),p) = pE_0 , \quad p > 1 , \qquad (6.44)$$

for some fixed value,

$$0 \le E_0 \le 1 . \tag{6.45}$$

When $p = 1$, the only choice is $E_0 = 1$. In other words, the speedup function is a linear function of image count with slope equal to a fixed fraction of perfect speedup.

To find how the problem size changes, recall definition (6.23) for the speedup function and definition (6.14) for the efficiency function, and apply the fixed-efficiency constraint (6.44) to obtain the formula,

$$E_0 = \frac{1}{1 + p/\gamma(n(p), p)} . \tag{6.46}$$

From definition (6.10) for the granularity function,

$$p/\gamma(n(p), p) = pt_{\text{sum}}(p)/t(n(p), 1) , \tag{6.47}$$

and under the assumption of constant power,

$$t(n(p), 1) = 2n(p)e_0/r_0 . \tag{6.48}$$

Hence,

$$pt_{\text{sum}}(p)/[2n(p)e_0/r_0] = (1 - E_0)/E_0 , \tag{6.49}$$

and the local partition size has the value,

$$n(p)/p = n(1)[E_0/(1 - E_0)]/\gamma(n(1), p) . \tag{6.50}$$

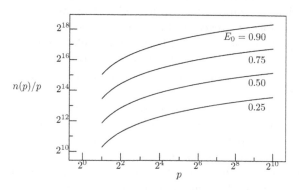

FIGURE 6.7: Partition size required to satisfy the fixed-efficiency constraint as a function of image count.

Figure 6.7 shows how the local partition size increases like $\log_2 p$ under the fixed-efficiency constraint. This growth in partition size puts a limit on the apparently unlimited scalability implied by Equation (6.44). Eventually, the partition size exceeds the available local memory.

If the algorithm is moved to a machine with higher computational power but the same synchronization time, the partition size increases to maintain the same efficiency. Similarly, if the synchronization time increases while the computational power remains the same, the partition size increases. Ideally, whenever a new processor design increases the computational power, it also decreases the synchronization time proportionally so that the algorithm runs at the same efficiency for the same partition size. Unfortunately, it is easier to increase computational power than it is to decrease synchronization time forcing the programmer to use larger partition sizes to maintain the same efficiency.

6.8 Some remarks on computer performance modeling

Computer performance modeling is the art of explaining measured execution time as a function of parameters that describe the software plus parameters that describe the hardware and also the art of explaining the interactions between them [52]. A good model gives the programmer an estimate of the execution time for a particular algorithm on a particular machine. If the execution time is too long, the programmer wants to know why and wants to know how to fix it. Even a good model, however, may not predict what to expect when an old machine is replaced by a new one.

Regrettably, the performance modeling community uses the term "scaling" to mean something different from the rest of the scientific community. Within the broader scientific community, scaling refers to scaling physical quantities to dimensionless form to reveal self-similarity properties [5] [7] [10]. Within the computer performance community, the term refers to how execution time behaves with respect to just one quantity, the processor count. This limited approach to performance modeling misses many important features hiding within a model [83] [84].

Table 6.1 compares four different constraints placed on the dot-product operation using model (6.4). For the strong-scaling case, the problem size remains fixed, the local partition size decreases rapidly with image count, and the decrease in execution time is severely limited. Under the fixed-time constraint, the problem size increases, the local partition size decreases, and the execution time remains the same. Under the fixed-work constraint, the problem size increases, the local partition size remains fixed, and the execution time increases. Under the fixed-efficiency constraint, the problem size increases, the local partition size increases, and the efficiency remains fixed.

There are other possible constraints that can be applied to define weak scaling [43] [44] [45] [80] [105]. Some performance reports, however, forget to state clearly what scaling constraint has been applied. Unfortunately, many years after Bailey wrote his satirical paper [3], understanding published performance results can still be difficult.

TABLE 6.1: Comparison of four constraints placed on the dot-product operation.

speedup: $s(n(p), p)$	partition size: $n(p)/p$	scaling constraint
$p/[1 + p(1 - \delta_1^p)/\gamma(n(1), p)]$	$n(1)/p$	fixed size
$p[1 - (1 - \delta_1^p)/\gamma(n(1), p)]$	$n(1)[1 - (1 - \delta_1^p)/\gamma(n(1), p)]$	fixed time
$p/[1 + (1 - \delta_1^p)/\gamma(n(1), p)]$	$n(1)$	fixed work
pE_0	$n(1)(E_0/(1 - E_0)/\gamma(n(1), p)$	fixed efficiency

Measuring execution time on modern computers is tricky [53]. Reproducing timing results from run to run, especially on very large machines, is almost impossible. The measured execution times for the sum-to-all operation represented by bullets in Figure 6.1 are average values. But as Figure 6.8 shows, there is a wide distribution of values around the average. These distributions are irregular with long tails, and the deviation from the average increases as the image count increases. Performance models often use average values to simplify the analysis, but the uncertainty in the measurements may be important on big machines.

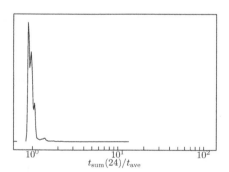
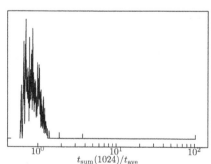

$t_{\text{sum}}(24)/t_{\text{ave}}$

$t_{\text{sum}}(1024)/t_{\text{ave}}$

FIGURE 6.8: Distribution of measured times for the sum-to-all operation, with $p = 24$ on the left and $p = 1024$ on the right, relative to the average time. In each case, the distribution represents a total of about one million measurements. The distribution broadens and flattens as the value of p increases.

There are many reasons for this uncertainty. As the image count increases, unpredictable interference occurs among images sharing common resources. Among other things, this interference depends on how programs are placed in the machine, what priorities they are assigned by the operating system, and the kind of communication patterns each program uses. The long tails in the distributions come from occasional system interrupts that can be several orders of magnitude longer than the execution time being measured. The uncertainty in measurement has been the subject of intense study as machines have grown in size, and the combined effects of the behavior have come to be called operating system jitter [6] [54] [103].

The reader may object that formula (6.1) omits the time required to perform the addition operation at each stage of the binary tree. A modern machine running at a frequency $\nu = 1$ GHz with synchronization time $\tau = 1$ μs can perform as many as $\tau\nu$ operations in the time it takes to synchronize. The addition operation, therefore, takes about three orders of magnitude less time than the synchronization time.

The reader may also question the assumption that the computational power is constant independent of problem size and image count. One answer to this legitimate question is that the model based on this assumption does a good job of representing the results for the dot-product operation. For other operations, the computational power may not be constant. Its behavior as a function of problem size and image count may be highly dependent on the details of a particular computer architecture. A detailed formula for the computational power may be complicated and hard to obtain without adding much new insight [41] [61] [62] [63] [64] [85]. Each algorithm must be analyzed on its own, but the basic techniques outlined here for the dot-product operation can still be used [43] [80] [105].

6.9 Exercises

1. Write code to measure the execution times for the sum-to-all function.

2. Write code to measure the execution times for the dot-product function.

3. Verify results (6.16) and (6.17).

4. Under what conditions are the two values for the minimum equal? What is the value of $p_{\min}$ in that case and what is the strong speedup?

5. Verify identities (6.32) and (6.33).

6. Verify results (6.40) and (6.41).

7. Calculate the position of the maximum in Figure 6.5 and the corresponding problem size at the maximum.

8. Verify result (6.50).

9. How does result (6.42) change if the computational power is not constant?

10. Suppose a fixed-time constraint uses the time on 32 processors. How does formula (6.34) change?

11. Consider an algorithm that performs $n^2 e_0$ floating-point operations followed by a global sum of the result. What is the weak speedup function under the constraint of fixed-partition size, $n(p) = pn(1)$?

Chapter 7

Partitioned Matrix Classes

This chapter combines the basic tools described in previous chapters to obtain more complicated applications than those already discussed. To facilitate the process, this book uses Fortran as a modern object-oriented language. The benefits of object-oriented design may not be apparent immediately to a traditional Fortran programmer. It is, however, an important topic to learn because many new parallel application codes use object-oriented techniques. Indeed, an exercise in object-oriented design often leads to a deeper understanding of parallel algorithms.

Matrices are important objects in many applications. Although matrix objects are represented as intrinsic two-dimensional arrays in Fortran, representing them as extensions of abstract matrix classes has many advantages. The abstract classes have data components common to all extensions without repeating them. Constructor functions create specific matrix objects with all the information contained within them that describes how the matrix is partitioned and how it is distributed across the machine. Interface blocks specify how procedures associated with type-bound procedure pointers must behave protecting the programmer from errors at compile-time. The programmer associates some of these procedure pointers to functions that translate indices from global to local and local to global and others to functions that perform linear algebra operations. The programmer can experiment with different implementations by associating pointers with different procedures residing in different libraries. The pointers change but the rest of an application code stays the same.

7.1 The abstract matrix class

Listing 7.1 shows the abstract matrix class. Since all matrices have a row dimension and a column dimension, the class includes data components that hold these values and all descendants of this class inherit these variables. The abstract class also includes data components for the number of images and the local image index. These quantities are not inherent properties of a matrix, but they occur so frequently in the definition of partitioned matrices and in associated procedures that it is convenient to include them as components of the abstract class. They also serve to record which image created an object and how many images were active at the time. All these variables are set to zero until a constructor function creates a specific matrix extension of the abstract class, and a destructor sets them back to zero when the object goes out of scope in a program.

Listing 7.1: The abstract matrix class.

```
module ClassAbstractMatrix
  implicit none
  type,abstract,public ::  AbstractMatrix
    integer             ::  p  = 0
    integer             ::  me = 0
    integer             ::  globalRowDim    = 0
    integer             ::  globalColumnDim = 0
  end Type AbstractMatrix
end module ClassAbstractMatrix
```

The reader familiar with object-oriented design might protest that the components of the abstract class should be private variables to prevent inadvertent corruption of their values. The only way, then, to set or retrieve their values would be through a collection of accessor functions. Although the use of accessor functions may be a good programming discipline, the clutter they induce tends to obfuscate the discussion of parallel techniques that are hard enough without other distractions.

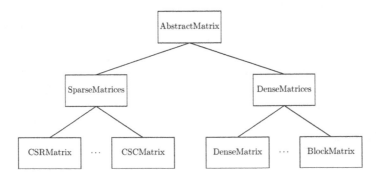

FIGURE 7.1: Inheritance from the abstract matrix class.

Figure 7.1 shows an outline of the plan for discussion of parallel linear algebra followed in the rest of the book. Two abstract classes, one for sparse matrices and one for dense matrices, are extensions to the base abstract matrix class. Sparse matrices have several different representations that extend the abstract sparse matrix class. The compressed-sparse-row (CSR) representation is, perhaps, most commonly used. Other possible extensions include the compressed-sparse-column (CSC) matrices, banded matrices, and multidiagonal matrices resulting from finite difference schemes.

Similarly, dense matrices have different representations depending on how they are blocked and how the blocks are assigned to images. The dense matrix class corresponds to a one-to-one map of blocks to images. The block matrix class corresponds to a many-to-one map. The blocks may even be permuted before they are assigned to images to form, for example, a block-cyclic matrix class.

7.2 Sparse matrix classes

Listing 7.2 shows a module that defines an abstract sparse matrix class that extends the basic abstract matrix class. It contains two data components, seven procedure pointers set to the null procedure by default, and interface blocks for each procedure. An integer component holds the number of nonzeros, an important quantity that desribes the structure of all sparse matrices. A parameter private to the module specifies double-precision variables for extensions of the class. Specifying a different value for this variable yields a single-precision version of the class.

Listing 7.2: An abstract class for sparse matrices.

```
module ClassSparseMatrices
  use ClassAbstractMatrix
  implicit none
  integer,parameter,private :: wp = kind(1.0d0)

  Type,abstract,extends(AbstractMatrix)   :: SparseMatrices
    integer                               :: nonZeros
    procedure(setSparsePointerInterface), &
                     nter :: setPointers => null()
    procedure(SparsePartitionInterface),  &
             pass(A),pointer :: partition => null()
    procedure(MatVecInterface),           &
                pass(A),pointer :: matVec => null()
    procedure(IterativeSolver),           &
                pass(A),pointer :: solve => null()
    procedure(ResidualInterface),         &
             pass(A),pointer :: residual => null()
    procedure(ReadMatrixInterface),       &
           pass(A),pointer :: readMatrix => null()
    procedure(WriteMatrixInterface),      &
          pass(A),pointer :: writeMatrix => null()
```

```
  end Type SparseMatrices

  abstract interface
    function MatVecInterface(A,x) result(y)
       import :: SparseMatrices, wp
       Class(SparseMatrices),intent(in) :: A
       real(wp),                 intent(in) :: x(:)
       real(wp),allocatable :: y(:)
    end function MatVecInterface
    function IterativeSolver(A,x,itMax,tolerance) result(y)
       import :: SparseMatrices, wp
       Class(SparseMatrices),intent(in) :: A
       real(wp),intent(in) :: x(:)
       integer, intent(in),optional :: itMax
       real(wp),intent(in),optional :: tolerance
       real(wp),allocatable  :: y(:)
    end function IterativeSolver
    function ResidualInterface(A,x,b) result(r)
       import :: SparseMatrices, wp
       Class(SparseMatrices),intent(in) :: A
       real(wp),intent(in) :: x(:),b(:)
       real(wp) :: r
    end function ResidualInterface
    subroutine ReadMatrixInterface(A,file)
       import :: SparseMatrices
       Class(SparseMatrices),intent(inout) :: A
       character(len=*),intent(in)     :: file
    end subroutine ReadMatrixInterface
    subroutine WriteMatrixInterface(A,file)
       import :: SparseMatrices
       Class(SparseMatrices),intent(in) :: A
       character(len=*),intent(in) :: file
    end subroutine WriteMatrixInterface
    subroutine setSparsePointerInterface(A)
       import :: SparseMatrices
       Class(SparseMatrices),intent(inout) :: A
    end subroutine setSparsePointerInterface
    subroutine SparsePartitionInterface(A)
       import :: SparseMatrices
       Class(SparseMatrices),intent(inout) :: A
    end subroutine SparsePartitionInterface
  end interface
end module ClassSparseMatrices
```

   One procedure pointer refers to a subroutine that sets the other pointers
to default procedures that reside in separate library modules. Another pointer
refers to a partition operator that partitions the sparse matrix in a way appropriate for the particular extension. The other procedure pointers refer to
matrix operations that should be familiar from their names. The matrix-vector

multiplication operation is required by almost all iterative solvers for sparse matrices, but the specific form taken by the operation depends on the specific representation of the matrix. Calculation of the residual is important for determining convergence of iterative solvers, and input-output operations provide access to sparse matrix data stored in files. The programmer may, of course, add other procedure pointers when defining extensions to the abstract class, and the programmer can override any or all of the default procedures as long as they conform to the interface blocks.

7.3   The compressed-sparse-row matrix class

The compressed-sparse-row (CSR) representation of a sparse matrix consists of three one-dimensional arrays [90, p. 89]. As shown in Figure 7.2, the array `csr(1:nnz)` holds the nonzero elements of the rows of the sparse matrix. The row-index array `ia(1:n+1)` holds indices corresponding to the start of each row such that `csr(ia(k))` is the first element of the k-th row. The column-index array `ja(1:nnz)` holds the column index for each element in the array `csr` such that element `csr(k)` belongs to column `ja(k)`.

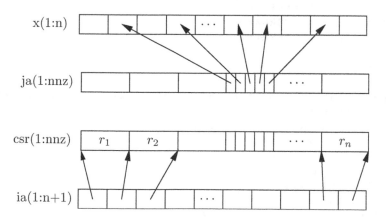

FIGURE 7.2: Compressed-sparse-row storage. The array `ia` holds indices pointing to the beginning of each row in the compressed data array `csr`. The array `ja` holds indices pointing to the corresponding column for each data element.

Listing 7.3 shows a module that defines the CSR sparse matrix class. It inherits all the information contained in its two parent abstract classes. The matrix is partitioned by rows, and the class contains five integer variables to describe the local partition: (1) the number of local rows, (2) the row remainder to account for non-uniform partitions, (3) the base index for the local

partition, (4) the local row dimension, and (5) the number of local nonzeros.
The constructor function sets the values for these variables, and it allocates
the arrays that describe the sparse structure to hold just the rows assigned to
the image.

Listing 7.3: The compressed-sparse-row (CSR) matrix class.

```fortran
module ClassCSRMatrix
  use ClassSparseMatrices
  implicit none
  integer,parameter,private :: wp=kind(1.0d0)
  Type,extends(SparseMatrices),public :: CSRMatrix
    integer :: rowPartitions = 0
    integer :: rowRemainder  = 0
    integer :: K_0           = 0
    integer :: localRowDim   = 0
    integer :: localNnz      = 0
    real(wp),allocatable :: csr(:)
    integer, allocatable :: ia(:)
    integer, allocatable :: ja(:)
  contains
    final :: deleteCSRMatrix
  end Type CSRMatrix

  interface CSRMatrix
    procedure newCSRMatrix
  end interface CSRMatrix

contains

  Type(CSRMatrix) function newCSRMatrix(n,nnz) result(A)
    implicit none
    integer,intent(in)  :: n,nnz
    procedure(setSParsePointerInterface) :: setCSRPointers
    A%p   = num_images()
    A%me  = this_image()
    A%globalRowDim    = n
    A%globalColumnDim = n
    A%nonZeros        = nnz
    A%setPointers     => setCSRPointers
    call A%setPointers()
    call A%partition()
  end function newCSRMatrix
  subroutine deleteCSRMatrix(A)
    Type(CSRMatrix) ::    A
    A%p               = 0
    A%me              = 0
    A%globalRowDim    = 0
    A%globalColumnDim = 0
    A%localRowDim     = 0
```

```
         A%rowRemainder     = 0
         A%rowPartitions    = 0
         A%K_0              = 0
         A%nonZeros         = 0
         A%localNnz         = 0
         A%solve           => null()
         A%readMatrix      => null()
         A%residual        => null()
         A%matVec          => null()
         A%setPointers     => null()
         A%partition       => null()
         if(allocated(A%csr)) deallocate(A%csr)
         if(allocated(A%ia))  deallocate(A%ia)
         if(allocated(A%ja))  deallocate(A%ja)
    end subroutine deleteCSRMatrix
end module ClassCSRMatrix
```

The constructor function accepts two integer arguments, the global matrix size and the number of nonzeros. It returns a CSR matrix object as shown in Listing 7.4.

Listing 7.4: Creating a CSR matrix object.

```
     use ClassCSRMatrix
     Type(CSRMatrix) :: A
            :
     A = CSRMatrix(n,nnz)
```

The constructor function records the values of the global dimensions and the number of nonzeros. It then assigns a procedure pointer to a subroutine, shown in Listing 7.5, that sets default values for the procedure pointers. The procedures are contained in the modules listed at the top of the subroutine. Since the procedure pointers are public variables, the programmer may assign them to other procedures. Reassigning procedure pointers can be dangerous, however, and the programmer must take care when doing so.

Listing 7.5: Code to set default values for CSR procedure pointers.

```
subroutine setCSRPointers(A)
  use ClassCSRMatrix
  use SparseLinearAlgebra, only : csrMatVec,residual
  use SparseIOLibrary,     only : read_CSR_matrix
  use ClassSolver,         only : cg
  use PartitionOperators,  only : partitionCSRMatrix
  implicit none
    Class(CSRMatrix),intent(inout) :: A
       A%matVec     => CSRMatVec
       A%solve      => cg
       A%residual   => residual
       A%readMatrix => read_CSR_Matrix
```

```
      A%writeMatrix => write_CSR_Matrix
      A%partition   => partitionCSRMatrix
end subroutine setCSRPointers
```

The constructor function also assigns a pointer to the partition operator shown in Listing 7.6. This procedure records the values of the variables that describe the row-partition operation applied to the matrix.

Listing 7.6: Partition operator for CSR matrices.

```
module PartitionOperators
 implicit none
 integer,parameter,private :: wp=kind(1.0d0)
contains
       :
 subroutine partitionCSRMatrix(A)
   use ClassCSRMatrix
   implicit none
   Class(CSRMatrix),intent(inout) :: A
   integer :: r
   A%rowPartitions = A%p
   r = mod(A%globalRowDim,A%rowPartitions)
   A%rowRemainder = r
   A%localRowDim   = (A%globalRowDim-1)/A%rowPartitions+1
   if(r/=0 .and. A%me>r) A%localRowDim = A%localRowDim-1
   A%K_0 = (A%me-1)*A%localRowDim
   if(A%me>r) A%K_0 = A%K_0+r
 end subroutine partitionCSRMatrix
       :
end module partitionOperators
```

The sparse matrix class also contains a destructor. The compiler automatically invokes the destructor whenever an object of this class goes out of scope. It provides a form of garbage collection that prevents memory leaks.

7.4   Matrix-vector multiplication for a CSR matrix

The constructor function for the CSR matrix class creates an object with a procedure pointer assigned to a default matrix-vector multiplication function like the one contained in the sparse linear algebra library module shown in Listing 7.7.

Listing 7.7: The sparse linear algebra library module.

```
module SparseLinearAlgebra
  use ClassCSRMatrix
```

```
      use Collectives, only : maxToAll, gatherToAll
      implicit none
      integer, parameter,private :: wp=kind(1.0d0)
      real(wp),parameter,private :: zero = 0.0_wp
contains
      function csrMatVec(A,x) result(y)
        implicit none
        Class(SparseMatrices),intent(in) :: A
        real(wp),intent(in)                :: x(:)
        real(wp),allocatable               :: y(:)
        real(wp),allocatable               :: z(:)
        integer :: i, j,k
        select type(A)
          class is (CSRMatrix)
          allocate(y(A%localRowDim))
          y = zero
          if(size(x)==A%localRowDim) then
            z = gatherToAll(x)
          else if(size(x)==A%globalColumnDim) then
            z = x
          end if
          do j=1,A%localRowDim
            do k=A%ia(j),A%ia(j+1)-1
              y(j) = y(j) + A%csr(k)*z(A%ja(k))
            end do
          end do
          class default
            write(*,"('Wrong Type for csrMatVec')")
            error stop
        end select
      end function csrMatVec

      function residual(A,x,b) result(r)
        Class(SparseMatrices),intent(in) :: A
        real(wp),intent(in)   :: x(:), b(:)
        real(wp) :: r
        real(wp),allocatable :: z(:)
          z = A%matVec(x) - b
          r = maxToAll(maxVal(abs(z)))
      end function residual
end module SparseLinearAlgebra
```

The function accepts two possibilities for the size of the incoming dummy array. If the incoming array has size equal to the global column dimension, it conforms with the column dimension of the CSR matrix. If its size equals the local row dimension, the function assembles the full vector using the gather-to-all function from the collectives library before performing the multiplication. The function always returns an array with size equal to the local row dimension. The library also contains other sparse functions such as the residual function shown at the bottom of Listing 7.7.

7.5   Exercises

1. Design a sparse matrix class using the alternative compressed-sparse-column (CSC) representation.

2. Write the matrix-vector multiplication function for your CSC matrix class.

3. Write a procedure that writes a CSR matrix to a file.

4. Design a tridiagonal matrix class.

Chapter 8

Iterative Solvers for Sparse Matrices

Iterative solvers for sparse systems of linear equations are perhaps the most important algorithms found in parallel application codes. Sparse matrices appear in finite difference and finite element methods for solving partial differential equations as well as in graph algorithms and artificial intelligence applications. The basic ingredients of these algorithms are matrix-vector multiplication and the dot-product operation. This chapter applies the fundamental techniques described in previous chapters to obtain a parallel conjugate gradient algorithm. Other Krylov-based algorithms involve the same fundamental operations, and conversion of other sparse solvers from sequential versions to parallel versions is straightforward after understanding the coversion of the conjugate gradient algorithm. The book does not discuss preconditioning, not because it is unimportant but because it is beyond the book's scope.

8.1   The conjugate gradient algorithm

The conjugate gradient algorithm for solving the system of equations,

$$Ax = b , \tag{8.1}$$

is an example of a Krylov subspace method [2, p. 23] [90, p. 190]. As shown in display (8.2), the algorithm uses three internal vectors, q, r, v, in addition to the right-hand-side vector b. Iterations continue until the norm of the residual $\sqrt{\alpha}$ is less than some tolerance ϵ. After convergence, the vector x holds the solution.

$$x = 0$$
$$r = b$$
$$q = r$$
$$\alpha = r \cdot r \qquad \text{dot product}$$
$$\texttt{iterate while}(\sqrt{\alpha} > \epsilon)$$
$$\quad v = Aq \qquad \text{matrix vector multiplication}$$
$$\quad \beta = q^T \cdot v \qquad \text{dot product} \qquad\qquad (8.2)$$
$$\quad \gamma = \alpha/\beta$$
$$\quad x = x + \gamma q \qquad \text{local axpy}$$
$$\quad r = r - \gamma v \qquad \text{local axpy}$$
$$\quad \alpha = r^T \cdot r \qquad \text{dot product}$$
$$\quad \gamma = \alpha/(\beta * \gamma)$$
$$\quad q = r + \gamma q \qquad \text{local axpy}$$
$$\texttt{end iterate}$$

Listing 8.1 shows a direct transcription of this algorithm as a Fortran function placed in a module. The function conforms to the interface for an iterative solver defined in Listing 7.2. Any sparse matrix object passed to this function as the first argument has a procedure pointer assigned to a sparse matrix-vector multiplication function needed to perform the algorithm. If the passed object is a CSR matrix, this pointer is associated with the function shown in Listing 7.7. The dot-product function, as described in Section 5.3, resides in the collectives module declared at the top of the solver module.

Listing 8.1: The sparse solver library.

```
module ClassSolver
 use ClassSparseMatrices
 use Collectives , only : dotproduct
 implicit none
 integer , parameter ,private  :: wp=kind (1.0d0)
 integer , parameter :: defaultMaxIterations = 100
 integer , private :: itCount
 real(wp),parameter ,private :: defaultTolerance=1.0e-09_wp
 real(wp),parameter ,private :: zero = 0.0_wp
 real(wp),parameter ,private :: one  = 1.0_wp
contains
    function cg(A,b,maxIterations ,tolerance) result(x)
       implicit none
       Class(SparseMatrices),intent(in) :: A
       real(wp),intent(in)            :: b(:)
       integer , optional ,intent(in)  :: maxIterations
       real(wp),optional ,intent(in)  :: tolerance
       real(wp),allocatable           :: x(:)
       real(wp),allocatable           :: r(:)
       real(wp),allocatable           :: p(:)
       real(wp),allocatable           :: v(:)
```

```
      real(wp)                         :: alpha
      real(wp)                         :: beta
      real(wp)                         :: eps
      real(wp)                         :: gamma
      integer                          :: itMax
      integer                          :: m
        if(present(maxIterations)) then
          itMax = maxIterations
        else
          itMax = defaultMaxIterations
        end if
        if(present(tolerance)) then
          eps = tolerance
        else
          eps = defaultTolerance
        end if
        m = size(b)
        allocate(x(m))
        allocate(r(m))
        allocate(p(m))
        allocate(v(m))
        x = zero
        r = b
        p = r
        alpha = dotproduct(r,r)
        itCount = 0
        do while(sqrt(alpha) > eps .and. itCount < itMax)
          itCount = itCount+1
          v         = A%matVec(p)
          beta      = dotproduct(v,p)
          gamma     = alpha/beta
          x         = x + gamma*p
          r         = r - gamma*v
          alpha     = dotproduct(r,r)
          gamma     = alpha/(gamma*beta)
          p         = r + gamma*p
        end do
      end function cg
!          :                  !
! Other iterative solvers   !
!          :                  !
end module ClassSolver
```

Listing 8.2 shows a program that uses the conjugate gradient algorithm to solve a system of equations for a CSR matrix. The program creates a CSR matrix by invoking the constructor function, which sets the pointer to the conjugate gradient solver and the pointer to the I/O procedure. It then reads data from the input file and allocates two arrays, one for the vector on the

right side and one vector for the solution. It fills the right-hand-side array
with data, invokes the solver, and computes the residual.

Listing 8.2: A program that uses the conjugate gradient solver.

```
program SolveIt
  use ClassCSRMatrix
  implicit none
  integer,parameter :: wp=kind(1.0d0)
  type(CSRMatrix) :: A
  real(wp),allocatable :: b(:),x(:)
  real(wp) :: resisual
  integer :: n, nnz
  character(len=7),parameter :: fileName = 'CSRMatrix'
    :
  A = CSRMatrix(n,nnz)
  call A%readMatrix(fileName)
  allocate(b(A%localRowDim))
  allocate(x(A%localRowDim))
  b    = someFunction()
  x    = A%solve(b)
  residual = A%residual(x,b)
end program SolveIt
```

Neither the code for the program nor the code for the iterative solver contains any co-array syntax. All co-array syntax is contained in library modules.
The programmer, by substituting other modules, can change the algorithms to
use a different programming model without changing the design of the sparse
matrix class and its extensions.

8.2   Other Krylov solvers

Other Krylov solvers are similar to the conjugate gradient solver. The
details change, but they all use matrix-vector multiplications and dot-product
operations. If the matrix is not symmetric positive definite, the programmer
can use the same matrix objects already defined and simply override the solver
by reassigning the procedure pointer as shown in Listing 8.3.

Listing 8.3: Reassigning the solver.

```
A%solver => biCGStab
```

The biconjugate gradient stabilized algorithm does not require a symmetric
positive definite matrix and is now the solver rather than the conjugate gradient algorithm.

8.3 Performance analysis for the conjugate gradient algorithm

The conjugate gradient algorithm is the basis for the High-Performance Conjugate Gradient benchmark referred to as the HPCG benchmark [30] [98]. Although the HPCG benchmark includes a preconditioner and uses a sparse matrix of a special form, it is still worth looking at a performance model for the simpler algorithm described in Section 8.1 because it illustrates how to think about analyzing algorithms like the conjugate gradient algorithm.

For the algorithm shown in Listing 8.1, a formula with four terms,

$$t(n, p) = t_{\text{matVec}}(n, p) + 3t_{\text{axpy}}(n, p) + t_{\text{gather}}(n, p) + 2t_{\text{dot}}(n, p) , \qquad (8.3)$$

is a good performance model for the execution time for a single iteration of the algorithm [84]. It is a function of the global matrix size n and the image count p.

Time is measured in seconds. The unit for work, $e_0 = 1$ flop, is the work for one 64-bit floating-point operation, and the unit of length, ℓ_0, is the length of one data element. Computational power, r_0, is the number of floating-point operations per second, measured in units of e_0/s. Bandwidth b_0 is the number of data elements moved per second, measured in units of ℓ_0/s, and τ_0 is the synchronization time, measured in seconds.

The matrix is square, symmetric, and positive definite and is partitioned by rows with partition size n/p the same for each image. With this partition size, the time for each local matrix-vector multiplication,

$$t_{\text{matVec}}(n, p) = 2\sigma \left(\frac{n}{p} \right) \left(\frac{e_0}{r_0} \right) , \qquad (8.4)$$

counting two floating-point operations for each nonzero entry σ in each of n/p rows. Although the number of nonzero entries per row depends on the particular sparse matrix and may change from row to row, assuming a constant number of nonzero entries detracts nothing important from the analysis.

The time for each of the three local 'axpy' operations obeys the formula,

$$t_{\text{axpy}}(n, p) = 2 \left(\frac{n}{p} \right) \left(\frac{e_0}{r_0} \right) , \qquad (8.5)$$

and the time for one dot-product operation obeys the formula,

$$t_{\text{dot}}(n, p) = 2 \left(\frac{n}{p} \right) \left(\frac{e_0}{r_0} \right) + 2\tau_0 \log_2 p , \qquad (8.6)$$

assuming there is just one physical processor per node so that the $k = 0$ term is the only term in formula (6.1).

The time for the gather operation obeys the formula,

$$t_{\text{gather}}(n, p) = (p - 1) \left(\frac{n}{p} \right) \left(\frac{\ell_0}{b_0} \right) , \tag{8.7}$$

where each image, to perform the matrix-vector multiplication operation, gathers vectors of length $(n/p)\ell_0$ from each of the other $(p-1)$ images with bandwidth b_0.

To continue the analysis, write the execution time per iteration as the sum of three terms,

$$t = t_1 + t_2 + t_3 , \tag{8.8}$$

by collecting together three terms from (8.4), (8.5), and (8.6) that refer to floating-point operations,

$$t_1 = 2(\sigma + 5) \left(\frac{n}{p} \right) \left(\frac{e_0}{r_0} \right) , \tag{8.9}$$

plus the term representing overhead from communication,

$$t_2 = (p - 1) \left(\frac{n}{p} \right) \left(\frac{\ell_0}{b_0} \right) , \tag{8.10}$$

plus the term for overhead from synchronization,

$$t_3 = 4\tau_0 \log_2 p . \tag{8.11}$$

With these definitions, the fraction of time spent in computation,

$$E = t_1/t , \tag{8.12}$$

is the efficiency function. It is a function of two dimensionless variables,

$$E(u, v) = \frac{1}{1 + u + v} , \tag{8.13}$$

where the two variables,

$$u = \left(\frac{r_0}{b_0} \right) \cdot \left(\frac{p - 1}{2(\sigma + 5)} \right) \cdot \left(\frac{\ell_0}{e_0} \right) , \tag{8.14}$$

$$v = \left(\frac{\tau_0 r_0}{\ell_0} \right) \cdot \left(\frac{p \log_2 p}{n(\sigma + 5)} \right) \cdot \left(\frac{\ell_0}{e_0} \right) , \tag{8.15}$$

are functions of problem size and image count. The variable $u = t_2/t_1$ measures the time for moving data relative to the time for computation, and the variable $v = t_3/t_1$ measures the time for synchronization relative to the time for computation.

These two variables, each a dimensionless ratio of times, are also dimensionless ratios of computational forces. A computational force is the ratio of work to length measured in units of flop/word or flop/byte, depending on the unit of length [77] [81] [82] [83]. Computational force is often called computational intensity [51] [52] [69] [71] or sometimes machine balance [14] [69]. The fundamental unit of force, e_0/ℓ_0, is the fundamental unit of work divided by the fundamental unit of length.

Two hardware forces appear in Equations (8.14) and (8.15),

$$f_B^H = \frac{r_0}{b_0} \tag{8.16}$$

$$f_L^H = \frac{\tau_0 r_0}{\ell_0} , \tag{8.17}$$

along with two software forces,

$$f_B^S = \left(\frac{2(\sigma+5)}{p-1}\right) \cdot \left(\frac{e_0}{\ell_0}\right) \tag{8.18}$$

$$f_L^S = \left(\frac{n(\sigma+5)}{p\log_2 p}\right) \cdot \left(\frac{e_0}{\ell_0}\right) . \tag{8.19}$$

The two variables defined by formulas (8.14) and (8.15), then, are ratios of these forces,

$$u = \frac{f_B^H}{f_B^S} \tag{8.20}$$

$$v = \frac{f_L^H}{f_L^S} . \tag{8.21}$$

As designated by subscripts, the first variable is related to bandwidth effects and the second variable is related to latency effects.

The first hardware force (8.16) represents the number of floating-point operations that could be done during the time it takes to transfer a single data element. The second hardware force (8.17) represents the number of floating-point operations that could be done during the time it takes to perform a synchronization across the machine. Two machines with different values for r_0, b_0 and τ_0, but with the same values for the two hardware forces, behave the same way when executing the conjugate gradient algorithm. Changing the three hardware parameters while keeping the hardware forces the same results in self-similar machines [5] [84]. The two software forces depend on the design of the conjugate gradient algorithm independent of hardware.

The efficiency function defines a surface above the two-dimensional (u, v) plane. As the problem size and image count change, these variables act like curvilinear coordinates that trace curves along the surface. Self-similar machines follow the same curve; dissimilar machines follow different curves. Traditional performance models draw the efficiency surface above the (n, p) plane rather than the (u, v) plane. This approach yields different surfaces for different machines and conflates bandwidth effects with latency effects. The (u, v)

plane defines a single surface for all machines and separates the bandwidth effects along one axis from the latency effects along the other axis.

The two coordinates are proportional to each other,

$$v = \left(\frac{\tau_0 b_0}{\ell_0}\right)\left(\frac{2p\log_2 p}{n(p-1)}\right)u \ . \tag{8.22}$$

The tangent vector at each point along a curve determines how fast the latency term increases relative to the bandwidth term. The point where the slope equals one,

$$dv/du = 1 \ , \tag{8.23}$$

determines a characteristic length,

$$\tau_0 b_0 = \left(\frac{n(p-1)}{2p\log_2 p}\right)\ell_0 \ . \tag{8.24}$$

Two machines with different values of τ_0 and b_0, but with the same product of the two values, reach this point at the same image count. This length is the ratio of the two hardware forces (8.16) and (8.17),

$$\frac{\tau_0 b_0}{\ell_0} = \frac{f_L^H}{f_B^H} \ . \tag{8.25}$$

The hardware forces, specific to a particular machine, interacting with the software forces, specific to a particular algorithm, determine all the properties of the curves along the surface. These observations dip a bit into the differential geometry of the efficiency surface. Further analysis of the surface, such as finding the geodesics, may result in new approaches to performance modeling [83].

8.4   Strong scaling

For the strong-scaling case, the matrix size n is fixed, and the software forces (8.18) and (8.19) are functions of just the image count p. The two software forces are the same for every machine. On the other hand, the two hardware forces (8.16) and (8.17) are different for different machines. Plotting the efficiency as a function of image count, therefore, yields different paths for different machines as shown in Figure 8.1.

A path along the surface represents a balancing act between software forces and hardware forces. The programmer must design algorithms to provide software forces larger than the hardware forces to obtain high efficiency. Indeed, if the two software forces just equal the two hadware forces, the efficiency is only one-third. The programmer's task is very difficult for the strong-scaling

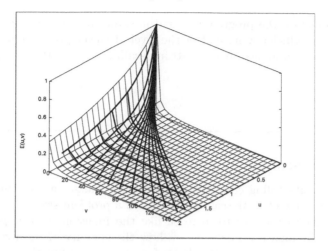

FIGURE 8.1: Paths along the efficiency surface for a fixed problem size $n = 1000$ as the image count changes. The first hardware force is fixed, $r_0/b_0 = 3.0$ flop/word. The second hardware force, $\tau_0 r_0/\ell_0$, changes value from 200 flop/word to 10000 flop/word in increments of 800 flop/word, defining different paths that start at the top of the surface at $E = 1$ for $p = 1$ and descend the surface as the image count increases. The curves across these paths mark the values of the efficiency when the image count is fixed starting with the value $p = 40$ at the top increasing to $p = 120$ in increments of 20.

case because the software forces go to zero as the image count increases and the two coordinates go to infinity,

$$\lim_{p \to \infty} u(n, p) = \infty , \quad \lim_{p \to \infty} v(n, p) = \infty , \quad n \text{ fixed} \qquad (8.26)$$

no matter the values of the hardware forces. The efficiency function, therefore, approaches zero,

$$\lim_{p \to \infty} E(u(n, p), v(n, p)) = 0 , \quad n \text{ fixed} . \qquad (8.27)$$

How quickly a path approaches zero and the direction it takes along the surface depend on the hardware forces.

It is possible, of course, to overcome hardware forces for a given machine by increasing the problem size for a fixed value of the image count. The efficiency function approaches one,

$$\lim_{n \to \infty} E(u_B(n, p), u_L(n, p)) = 1 , \quad p \text{ fixed} , \qquad (8.28)$$

because the two software forces increase with problem size eventually overcoming the hardware forces. Low efficiency on a given machine, however, may suggest that the algorithm needs to change such that the software forces increase enough to obtain acceptable efficiency. This alternative, however, places a heavy burden on the programmer, and in some cases it may be impossible to obtain good efficiency no matter the level of clever programming skill. In most cases, programmers abandon strong scaling and resort to some form of weak scaling.

8.5 Weak scaling

For the weak-scaling case, the problem size changes as the image count changes. One form of weak scaling starts with a problem size $n(1)$ on a single image and allows the size to increase like the function $n(p) = pn(1)$. The behavior of the algorithm depends on how the sparsity of the rows changes as the image count changes. One choice for the sparsity function is to assume that it changes in the same way as the problem size changes, $\sigma(p) = p\sigma_0$ where σ_0 is the sparsity of the original matrix. The two curvilinear coordinates, then, become functions of just the image count,

$$u(p) = \left(\frac{r_0}{b_0}\right) \cdot \left(\frac{p-1}{2(p\sigma_0 + 5)}\right) \cdot \left(\frac{\ell_0}{e_0}\right) , \tag{8.29}$$

$$v(p) = \left(\frac{\tau_0 r_0}{\ell_0}\right) \cdot \left(\frac{p \log_2 p}{pn(1)(p\sigma_0 + 5)}\right) \cdot \left(\frac{\ell_0}{e_0}\right) . \tag{8.30}$$

with n replaced with $pn(1)$ and σ replaced with $p\sigma_0$.

In the limit of large p, these functions behave quite differently from the strong-scaling case. The latency term goes to zero,

$$\lim_{p\to\infty} v(p) = 0 , \tag{8.31}$$

while the bandwidth term approaches a finite limit,

$$\lim_{p\to\infty} u(p) = \left(\frac{1}{2\sigma_0}\right)\left(\frac{r_0\ell_0}{b_0 e_0}\right) . \tag{8.32}$$

The efficiency function approaches the limit,

$$\lim_{p\to\infty} E(p) = \frac{1}{1 + (r_0\ell_0)/(2\sigma_0 b_0 e_0)} . \tag{8.33}$$

It is important to notice, however, that if the row sparsity remains constant as the problem size increases, rather than growing with problem size, both coordinates approach infinity for large image counts.

Figure 8.2 shows the paths followed for different machines for the weak-scaling case. These paths are quite different from those shown in Figure 8.1 for the strong-scaling case. The paths for the weak case turn around at some

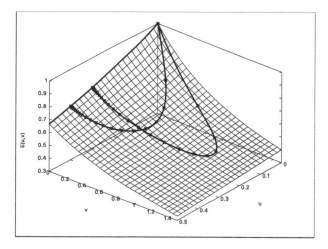

FIGURE 8.2: Paths along the efficiency surface for the weak-scaling case. The initial problem size is $n(1) = 100$ with row sparsity $\sigma_0 = 10$. The points along the paths mark the increase in image count from one at the pinnacle of the surface to very large values as the paths approach their asymptotic values. The path that descends lowest on the surface corresponds to a machine with hardware parameters $r_0/b_0 = 6$ and $\tau_0 r_0 = 2500$. The other path corresponds to a machine with parameters $r_0/b_0 = 8$ and $\tau_0 r_0 = 1200$.

value of p and approach the limits given by (8.31)-(8.33). Following along the path as p increases, the turning point occurs when the derivative of the efficiency function goes to zero,

$$dE/dp = -(1/E^2)(du/dp + dv/dp) = 0 \ . \qquad (8.34)$$

This derivative is zero if and only if the derivative of one curvilinear coordinate equals the negative derivative of the other coordinate,

$$du/dp = -dv/dp \ , \qquad (8.35)$$

or, in other words, if the derivative of one coordinate with respect to the other obeys the relationship,

$$dv/du = -1 \ . \qquad (8.36)$$

This point provides another characteristic length different from the previous length determined by (8.23).

The projections of these paths onto the underlying (u, v) plane are suggestive of phase portraits for some differential equation with an asymptotic attractive node. It is likely a second order equation with independent variable p, but what differential equation it might be is an interesting question yet to be answered.'

8.6   Iso-efficiency

Another form of weak scaling results from the iso-efficiency constraint [67]. The problem size is a function of image count determined in such a way that the efficiency remains constant,

$$E_0 = \frac{1}{1 + u(n(p), p) + v(n(p), p)} \ . \tag{8.37}$$

This equation defines a path along the surface at $E = E_0$ such that

$$u(n(p), p) + v(n(p), p) = \frac{1 - E_0}{E_0} \ . \tag{8.38}$$

This path projects to a straight line in the (u, v) plane, and the function $n(p)$ can be found graphically [81].

8.7   Exercises

1. Design a column-partitioned matrix class such that the conjugate gradient code works without change.

2. Pick another Krylov solver, for example the biCGStab solver [90, p. 234], and convert it to a partitioned algorithm.

3. Verify formulas (8.14) and (8.15).

4. Verify formula (8.22).

5. Determine the asymptotic behavior of the coordinates (u, v) as the image count increases under the assumption that the row sparsity remains constant.

6. Find the characteristic length implied by formula (8.36).

7. Sketch the projection of the paths shown in Figure 8.2 onto the (u, v) plane. Where do the maximum values occur?

Chapter 9

Blocked Matrices

A partition operator applied to the rows of a matrix from the left and a reverse partition operator applied to the columns of a matrix from the right yield a blocked matrix. As long as the number of row partitions times the number of column partitions equals the number of images, the map between matrix blocks and image indices is a one-to-one map. The single image index becomes two indices within a logical two-dimensional grid, and each block is labeled with two co-dimension indices within the grid.

Parallel algorithms for dense linear algebra are less important now than they once were, but they were among the first parallel algorithms implemented and it is important to understand the issues involved in their design. Indeed, much of the co-array model originated as a natural extension from array syntax to co-array syntax that provided a description of blocked data structures distributed across a parallel machine [78] [79].

After Fortran became an object-oriented language, the co-array model fit well into the design of distributed classes. The dense matrix class described in this chapter helps the programmer deal with perhaps the most difficult task involved in the design of parallel algorithms, the manipulation of maps between local and global indices. The class contains all the information about these maps along with procedure pointers to basic linear algebra operations. Its design resembles the design of the sparse matrix class described in Chapter 7.

9.1 Partitioned dense matrices

Application of the partition of unity (4.3) on both the left and the right of a matrix,

$$A = \left(\sum_{\alpha=1}^{q} P_\alpha P^\alpha \right) A \left(\sum_{\beta=1}^{r} P_\beta P^\beta \right) , \tag{9.1}$$

yields a double sum,

$$A = \sum_{\alpha=1}^{q} \sum_{\beta=1}^{r} P_\alpha A_\beta^\alpha P^\beta , \tag{9.2}$$

over partitions of the matrix,

$$A_\beta^\alpha = P^\alpha A P_\beta , \quad \alpha = 1,\ldots,q , \quad \beta = 1,\ldots,r . \tag{9.3}$$

Each partition is called a block and the number of blocks must equal the number of images,

$$p = q \times r , \tag{9.4}$$

arranged in a $[q,r]$ logical grid. This constraint is not strictly necessary, but matrix algorithms are more complicated if the map is not one-to-one. For now, the discussion assumes a one-to-one map in order not to raise issues that just tend to hide the intrinsic nature of parallel algorithms for blocked matrices. It is often necessary, however, to lift this constraint to achieve good performance, and Chapter 12 describes these more complicated algorithms.

9.2 An abstract class for dense matrices

The abstract class for dense matrices, shown in Listing 9.1, is an extension of the abstract matrix class corresponding to the box on the right side of Figure 7.1. It consists of a set of procedure pointers for basic linear algebra operations commonly applied to dense matrices. When the constructor function creates a dense matrix object as an extension of this abstract class, it associates these pointers with procedures that conform to the interfaces defined by the abstract class. The programmer can add other procedure pointers to the abstract class or to specific extensions of the abstract class.

Listing 9.1: The abstract dense matrices class.

```
module ClassDenseMatrices
  use ClassAbstractMatrix
  implicit none
```

```
      integer,parameter,private :: wp = kind(1.0d0)

   Type,abstract,extends(AbstractMatrix) :: DenseMatrices
      procedure(DenseMatMulInterface),          &
                  pass(A),pointer :: matMul        => null()
      procedure(DenseMatVecInterface),          &
                  pass(A),pointer :: matVec        => null()
      procedure(DenseTransposeInterface),       &
                  pass(A),pointer :: transpose     => null()
      procedure(DenseLUInterface),              &
                  pass(A),pointer :: LU            => null()
      procedure(DenseLUSolveInterface),         &
                  pass(A),pointer :: LUSolve       => null()
      procedure(DenseExchangeInterface),        &
                  pass(A),pointer :: haloExchange => null()
      procedure(DenseResidualInterface),        &
                  pass(A),pointer :: residual      => null()
      procedure(readDenseMatrix),               &
                  pass(A),pointer :: readMatrix    => null()
      procedure(writeDenseMatrix),              &
                  pass(A),pointer :: writeMatrix   => null()
      procedure(setPointerInterface),           &
                  pass(A),pointer :: setPointers   => null()
      procedure(DensePartitionInterface),       &
                  pass(A),pointer :: partition     => null()
   end Type DenseMatrices

   abstract interface
                  :
         ! abstract interface blocks !
                  :
   end interface
end module ClassDenseMatrices
```

9.3   The dense matrix class

Listing 9.2 shows the dense matrix class, an extension to the abstract dense matrix class. An object of this class allocates a two-dimensional array component to hold data for the local block assigned to the image that creates the object. The class also contains an array to hold pivot information for the LU decomposition algorithm as described in Section 9.5. Other data components hold information describing the partition operations for rows and columns.

Listing 9.2: The dense matrix class.

```fortran
module ClassDenseMatrix
use ClassDenseMatrices
implicit none
integer,parameter,private   :: wp=kind(1.0d0)
Type,extends(DenseMatrices) :: DenseMatrix
 real(wp),allocatable       :: a(:,:)
 integer, allocatable       :: pivot(:,:)
 integer :: coDim1           = 0
 integer :: coDim2           = 0
 integer :: K_0              = 0
 integer :: L_0              = 0
 integer :: myP              = 0
 integer :: myQ              = 0
 integer :: rowRemainder     = 0
 integer :: columnRemainder  = 0
 integer :: localRowDim      = 0
 integer :: localColumnDim   = 0
 integer :: maxDims(2)       = 0
 integer :: haloWidth        = 0

contains
 final :: deleteDenseMatrix
end Type DenseMatrix

interface DenseMatrix
  procedure newDenseMatrix
end interface DenseMatrix

contains
 Type(DenseMatrix) &
   function newDenseMatrix(m,n,coDim1,coDim2,haloWidth) &
   result(A)
   integer,intent(in) :: m,n
   integer,intent(in) :: coDim1
   integer,intent(in) :: coDim2
   integer,intent(in),optional :: haloWidth
   integer            :: r
   procedure(setPointerInterface) :: setDensePointers
    A%p  = num_images()
    A%me = this_image()
    if(coDim1*coDim2 /= A%p) then
     write(*,"('Error in newDenseMatrix')")
     write(*,"('coDim1 x coDim2 /= p')")
     error stop
    end if
    A%coDim1 = coDim1
    A%coDim2 = coDim2
    A%globalRowDim    = m
    A%globalColumnDim = n
```

```
      if(present(haloWidth)) A%haloWidth = haloWidth
      A%setPointers => setDensePointers
      call A%setPointers()
      call A%partition()
   end function newDenseMatrix
            :
!...code for the final procedure...!
            :
end module ClassDenseMatrix
```

The constructor function accepts up to five arguments. The first two are the global row and column dimensions of the matrix; the second two are the co-dimensions used to partition the matrix, the first for the rows and the second for the columns. The fifth argument is optional and set to zero by default. If the programmer supplies a nonzero value, a halo of extra elements surrounds each block to support, for example, algorithms for solving finite difference equations. After recording the information supplied by its arguments and checking that the product of the two co-dimensions equals the number of images, the constructor sets a pointer to the procedure shown in Listing 9.3 that associates the procedure pointers to default procedures.

Listing 9.3: Code for setting default procedure pointers.

```
subroutine setDensePointers(A)
  use ClassDenseMatrix
  use DenLinAlg, only : denseMatVec, denseMatMul,      &
                        denseTranspose, denseLU,       &
                        denseLUSolve, denseExchange,   &
                        denseResidual
  use DenseIOLibrary,     only : read_Dense_Matrix
  use PartitionOperators, only : partitionDenseMatrix
  implicit none
  Class(DenseMatrix),intent(inout) :: A
    A%matVec       => denseMatVec
    A%matMul       => denseMatMul
    A%transpose    => denseTranspose
    A%LU           => denseLU
    A%LUSolve      => denseLUSolve
    A%haloExchange => denseExchange
    A%residual     => denseResidual
    A%readMatrix   => read_Dense_Matrix
    A%writeMatrix  => write_Dense_Matrix
    A%partition    => partitionDenseMatrix
end subroutine setDensePointers
```

The constructor associates one of the pointers to the procedure shown in Listing 9.4. It fills in the information that defines how the matrix has been partitioned and assigned to images. It also allocates the array component with appropriate local dimensions and initializes the array to zero.

Listing 9.4: Code to partition a dense matrix.

```
module PartitionOperators
 implicit none
 integer,parameter,private :: wp=kind(1.0d0)
contains
 subroutine partitionDenseMatrix(A)
  use ClassDenseMatrix
  implicit none
  Class(DenseMatrix),intent(inout) :: A
  integer :: r,m,n,w
   A%myQ = (A%me-1)/A%coDim1 + 1
   A%myP = A%me - (A%myQ-1)*A%coDim1
   A%maxDims(1) = (A%globalRowDim-1)/A%coDim1 + 1
   A%maxDims(2) = (A%globalColumnDim-1)/A%coDim2 + 1
   r = mod(A%globalRowDim,A%coDim1)
   A%rowRemainder = r
   A%localRowDim  = A%maxDims(1)
   if(r/=0 .and. A%myP>r) A%localRowDim = A%maxDims(1)-1
   A%K_0 = (A%myP-1)*A%localRowDim
   if(A%myP>r) A%K_0 = A%K_0+r
   r = mod(A%globalColumnDim,A%coDim2)
   A%columnRemainder = r
   A%localColumnDim  = A%maxDims(2)
   if(r/=0 .and. A%myQ>r) A%localColumnDim=A%maxDims(2)-1
   A%L_0 = (A%myQ-1)*A%localColumnDim
   if(A%myQ>r) A%L_0 = A%L_0+r
   m=A%localRowDim
   n=A%localColumnDim
   w=A%haloWidth
   allocate(A%a(1-w:m+w,1-w:n+w))
   A%a = 0.0_wp
 end subroutine partitionDenseMatrix
          :
end module partitionOperators
```

9.4   Matrix-matrix multiplication

Matrix-matrix multiplication,

$$C = AB , \tag{9.5}$$

with the insertion of the partition of unity in appropriate places,

$$\sum_{\alpha=1}^{q}\sum_{\beta=1}^{r} P_\alpha\left[P^\alpha C P_\beta\right] P^\beta = \sum_{\alpha=1}^{q}\sum_{\beta=1}^{r} P_\alpha\left[\sum_{\gamma=1}^{s}(P^\alpha A P_\gamma)(P^\gamma B P_\beta)\right] P^\beta , \tag{9.6}$$

defines a partitioned algorithm where the image with co-dimension indices $[\alpha, \beta]$ computes the block it owns,

$$C_\beta^\alpha = \sum_{\gamma=1}^{s} A_\gamma^\alpha B_\beta^\gamma , \quad \alpha = 1,\ldots,q , \quad \beta = 1,\ldots,r . \qquad (9.7)$$

The sum over the index γ indicates that each image must obtain blocks of the first matrix on the right side across the row corresponding to its first co-dimension index and blocks of the second matrix down the column corresponding to its second co-dimension index. It multiplies the blocks together and sums the results as shown in Figure 9.1 [37].

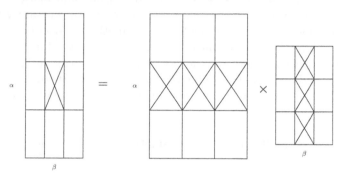

FIGURE 9.1: Partitioned matrix-matrix multiplication on nine images. To compute the block marked by a cross on the left, the image owning the block must obtain all the blocks marked by crosses across a row of the image grid from the first matrix on the right and all the blocks marked by crosses down a column of the image grid from the second matrix.

The conformal rules for matrix multiplication require the column dimension of matrix A to equal the row dimension of matrix B. The one-to-one assignment of blocks to images requires the product of the number of blocks for the outer two partitions to equal the number of images,

$$p = qr , \qquad (9.8)$$

such that each image owns just one block of the C matrix. In addition, the number of blocks, s, for the inner partition of unity must satisfy two constraints,

$$qs = p , \quad sr = p , \qquad (9.9)$$

such that each image owns just one block of each of the other two matrices. Independent of the value, s, the two constraints reduce to a single constraint,

$$q = r , \qquad (9.10)$$

and from (9.8), the number of images must be square,

$$p = q^2 . \qquad (9.11)$$

Furthermore, since $q^2 = qr$, the inner partition size must also have the value $s = q$. All three matrices, therefore, must be partitioned into a square image grid. The matrices themselves need not be square, but the logical image grid must be square.

When these constraints are satisfied, matrix-matrix multiplication is fairly simple to write as shown in Listing 9.5. Each image computes the result for the block assigned to it, marked with a cross on the left side of Figure 9.1, by accumulating results obtained by multiplying blocks assigned to other images along the same row of images, also marked with crosses, times blocks assigned to other images down the same column.

The function creates the result matrix object using information contained in the two input matrices. It also allocates two temporary co-array variables that it uses to move blocks from one image to another. It inserts these blocks as arguments into the local intrinsic matrix-multiplication function, and the compiler generates code to obtain the blocks from the images designated by the co-dimension indices. The two outer co-dimension indices correspond to the indices of the invoking image, and the sum traverses the middle co-dimension index.

Listing 9.5: Dense matrix-matrix multiplication.

```
module DenLinAlg
  use ClassDenseMatrix
  use collectives, only : maxToAll, sumToAll, gatherToAll
  implicit none
  integer,parameter,private  :: wp   = kind(1.0d0)
  integer,parameter,private  :: zero = 0.0_wp
contains


          :

  Type(DenseMatrix) function denseMatMul(A,B) result(C)
    Class(DenseMatrix),intent(in) :: A,B
    integer :: p,dimA(2),dimB(2)
    integer :: k,L,m,n,r
    real(wp),allocatable :: tempA(:,:)[:,:]
    real(wp),allocatable :: tempB(:,:)[:,:]
    dimA = A%maxDims
    dimB = B%maxDims
    C = DenseMatrix(                                        &
                    A%globalRowDim,B%globalColumnDim,  &
                    A%coDim1,B%coDim2,A%haloWidth)
    allocate(tempA(dimA(1),dimA(2))[A%coDim1,*])
    allocate(tempB(dimB(1),dimB(2))[B%coDim1,*])
    k = A%localRowDim
    L = A%localColumnDim
    tempA(1:k,1:L) = A%a(1:k,1:L)
    k = B%localRowDim
```

```
    L = B%localColumnDim
    tempB(1:k,1:L) = B%a(1:k,1:L)
    m=C%localRowDim
    n=C%localColumnDim
    k=dimA(2)
    r=A%columnRemainder
    sync all
    do p=1,A%coDim2
      C%a(1:m,1:n) = C%a(1:m,1:n)
                    + matmul(tempA(1:m,1:k)[A%myP,p], &
                             tempB(1:k,1:n)[p,B%myQ])
      if(p==r) k=k-1
    end do
    deallocate(tempA)
    deallocate(tempB)
  end function denseMatMul

        :

end module DenLinAlg
```

Other versions for matrix-matrix multiplication result from inserting the partition of unity in different ways [79]. Each version has its place in the appropriate circumstances.

$$
\begin{aligned}
C &= \sum_{\gamma} A_{\gamma} B^{\gamma} \\
C^{\alpha} &= \sum_{\gamma} A_{\gamma}^{\alpha} B^{\gamma} \\
C_{\beta} &= \sum_{\gamma} A_{\gamma} B_{\beta}^{\gamma} \\
C_{\beta}^{\alpha} &= \sum_{\gamma} A_{\gamma}^{\alpha} B_{\beta}^{\gamma} \\
C_{\beta}^{\alpha} &= A^{\alpha} B_{\beta} \\
C^{\alpha} &= A^{\alpha} B \\
C_{\beta} &= A B_{\beta}
\end{aligned}
\tag{9.12}
$$

9.5 LU decomposition

The LU-decomposition algorithm is a popular method for solving dense systems of linear equations,

$$
Ax = b \ . \tag{9.13}
$$

It factors the matrix into the product of a lower triangular matrix L and an upper triangular matrix U,

$$A = LU \; , \tag{9.14}$$

such that finding the solution is a two-step process, first solving the lower-triangular system,

$$Ly = b \; , \tag{9.15}$$

using forward substitution, followed by solving the upper-triangular system,

$$Ux = y \; , \tag{9.16}$$

using backward substitution.

Listing 9.6: Sequential LU decomposition.

```
do k=1,n
    a(k+1:n,k) = a(k+1:n,k)/a(k,k)
    do j=k+1,n
        a(k+1:n,j) = a(k+1:n,j) - a(k+1:n,k)*a(k,j)
    end do
end do
```

Code for the sequential algorithm assumes the simple form shown in Listing 9.6. As shown in Figure 9.2, the algorithm starts at the first element in the upper left corner and proceeds down the diagonal. At each step, it scales the column below the diagonal by the diagonal element and modifies the submatrix to the right and down by rank-one updates using the scaled column below the diagonal and the row to the right of the diagonal. Since the global matrix resides in local memory, implementing the algorithm requires manipulation of normal array indices only to reference data as needed.

The parallel implementation follows the same logic, but some of the data required at each step is remote data. To see how to implement the parallel algorithm, insert the partition of unity at appropriate places in Equations (9.13) and (9.14),

$$\sum_{\alpha=1}^{q}\sum_{\gamma=1}^{q}\sum_{\beta=1}^{q} P_\alpha (P^\alpha L P_\gamma)(P^\gamma U P_\beta) P^\beta x = \sum_{\alpha=1}^{q} P_\alpha P^\alpha b \; , \tag{9.17}$$

to obtain the system of equations in blocked form,

$$\sum_{\gamma=1}^{q}\sum_{\beta=1}^{q} L^\alpha_\gamma U^\gamma_\beta x^\beta = b^\alpha \; , \quad \alpha = 1, \ldots, q \; . \tag{9.18}$$

To maintain a one-to-one map between block indices and image indices, constraint (9.11) implies that the image count is square just as it did for matrix multiplication. The diagonal blocks, therefore, are labeled by co-dimension indices, [1,1], [2,2], ..., [A%coDim1,A%coDim1].

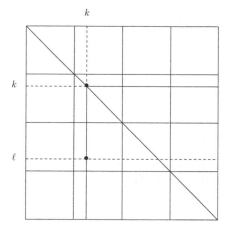

FIGURE 9.2: LU decomposition. The algorithm works down the diagonal reducing data below and to the right. In the parallel case, the matrix is partitioned into blocks, and the image owning the diagonal block controls the flow of the algorithm. Images below and to the right must wait to obtain data from the image on the diagonal. In addition, a pivot row below the diagonal row may need to be swapped with the diagonal row.

Listing 9.7: LU decomposition for dense matrices.

```
module DenLinAlg
  use ClassDenseMatrix
  use collectives, only : maxToAll, sumToAll, gatherToAll
  implicit none
  integer,parameter,private   :: wp   = kind(1.0d0)
  integer,parameter,private   :: zero = 0.0_wp
contains
                :
function denseLU(A) result(LU)
    Class(DenseMatrices),intent(in)  :: A
    Type(DenseMatrix),allocatable    :: LU
    integer,allocatable              :: iPivot(:,:)
    real(wp),allocatable :: swap(:)[:,:]
    real(wp),allocatable :: tempRow(:)[:,:]
    real(wp),allocatable :: tempCol(:)[:,:]
    real(wp),allocatable :: pivotElement(:)[:,:]
    integer, allocatable :: iPiv(:,:)[:,:]
    integer :: i,j,k,i0,j0
    integer :: m,n,p,q,r
    integer :: myP,myQ
    integer :: B,mB,nB

  select type(A)
    class is (DenseMatrix)
```

```
        mB = A%maxDims(1)
        nB = A%maxDims(2)
        p = A%coDim1
        q = A%coDim2
        myP = A%myP
        myQ = A%myQ
        m = A%localRowDim
        n = A%localColumnDim
        r = A%columnRemainder

        allocate(tempRow(nB)[p,*])
        allocate(tempCol(mB)[p,*])
          :
        LU = A

        do B=1,p
          do k=1,nB
#ifdef PIVOT
          :
#endif
          if(myP==B .and. myQ==B) then
             LU%a(k+1:m,k) = LU%a(k+1:m,k)/LU%a(k,k)
             tempCol(k:m) = LU%a(k:m,k)
             tempRow(k:n) = LU%a(k,k:n)
             i0=k+1
             j0=k+1
           end if

           sync all

           if(myP==B .and. myQ>B) then
             tempRow(1:n) = LU%a(k,1:n)
             tempCol(k:m) = tempCol(k:m)[B,B]
             i0=k+1
             j0=1
           end if

           if(myP>B .and. myQ==B) then
             tempRow(k:n) = tempRow(k:n)[B,B]
             LU%a(1:m,k) = LU%a(1:m,k)/tempRow(k)
             tempCol(1:m) = LU%a(1:m,k)
             i0=1
             j0=k+1
           end if

           sync all

           if(myP>B .and. myQ>B) then
             tempRow(1:n) = tempRow(1:n)[B,myQ]
```

```
            tempCol(1:m) = tempCol(1:m)[myP,B]
            i0=1
            j0=1
          end if

        if(myP>=B .and. myQ>=B) then
          do j=j0,n
            LU%a(i0:m,j) = LU%a(i0:m,j) &
                           - tempCol(i0:m)*tempRow(j)
          end do
        end if

      end do

    if(B==r) nB=nB-1
    sync all

    end do
      :
  end select
end function denseLU
      :
end module DenLinAlg
```

Listing 9.7 shows a module that contains a condensed version of a function that performs the parallel LU-decomposition algorithm. It ignores for now the pivot operation discussed later in Section 9.6. Before starting the decomposition, each image copies its local input matrix into the return matrix with automatic allocation of the return object. At the end of the function, the return matrix contains the matrix in factored form while the original matrix remains unchanged.

The logic of the parallel algorithm is the same as the logic for the sequential algorithm with elements replaced by blocks. The algorithm works down the diagonal blocks. The image whose co-dimension indices equal the block indices, [myP=B,myQ=B], owns the block and works its way down the diagonal of the block, k=1,...,nB. The diagonal image scales the elements below the diagonal element and places the current row and current column into temporary co-array variables to make them visible to other images. The two indices, i0=k+1,j0=k+1, are the indices of the submatrix to be reduced.

A synchronization point is required before other images access data from the diagonal image. All the images that own blocks in the same row as the diagonal image but to the right of the diagonal, that is with co-dimension indices [myP=B,myQ>B], obtain the temporary column vector from the diagonal image. Each of these images fills its temporary row vector with values needed by images in the same column but below the diagonal. The submatrix that needs to be reduced by these images starts at element i0=k+1, j0=1.

Images in the same column as the diagonal image but below the diagonal, that is with co-dimension indices [myP>B,myQ=B], obtain the temporary row vector from the diagonal image and fill a temporary column vector for images in the same row but to the right of the diagonal. The submatrix that needs to be reduced by these images starts at element i0=1, j0=k+1.

Images below the diagonal and to the right, that is with co-dimension indices [myP>B,myQ>B], must wait at a synchronization point before obtaining a temporary column vector from the image to the left and a temporary row vector from the image above. The submatrix that needs to be reduced for these processors starts at element i0=1, j0=1.

Finally, images with co-dimension indices [myP>=B,myQ>=B], on the diagonal, below the diagonal, and to the right of the diagonal, perform rank-one updates of their blocks. Images adjust the block size as needed and wait at a synchronization point before moving to next diagonal block.

This version of the LU-decomposition algorithm does not scale well because images drop out of work as they descend down the diagonal blocks. It illustrates how co-dimension indices map directly onto blocked linear algebra operations, but it does not produce a good parallel algorithm. Section 13.3 describes a better implementation using partition operators that do not assume a one-to-one map between images and blocks.

9.6   Partial pivoting

The LU-decomposition algorithm is unstable except for special cases [29] [101]. The usual fix for this problem is to permute the rows of the matrix such that at each stage in the algorithm the element with the largest absolute value in the current column is moved to the diagonal. In Figure 9.2, for example, row ℓ swaps with row k. As shown in Listing 9.8 for the sequential algorithm, this pivot operation occurs before scaling elements below the diagonal.

Listing 9.8: Partial pivoting for sequential code.

```
do k=1,n
    pivot(k) = k
    j=maxLoc(abs(a(k:n,k)),1)+k-1
    if(j/=k) then
        temp(1:n) = a(k,1:n)
        a(k,1:n)  = a(j,1:n)
        a(j,1:n)  = temp(1:n)
        pivot(k)  = j
    end if
    a(k+1:n,k)  = a(k+1:n,k)/a(k,k)
    do j=k+1,n
        a(k+1:n,j) = a(k+1:n,j) - a(k+1:n,k)*a(k,j)
    end do
```

```
      end do
```

For the parallel algorithm, the pivot operation is necessarily more complicated. The search for the pivot requires a search across images on or below
the diagonal image, and the row exchange involves all the images in the same
row as the diagonal image and all images in the same row as the image that
holds the pivot row.

Listing 9.9: Partial pivoting for parallel code.

```
        allocate(iPivot(2,m))
        allocate(swap(nB)[p,*])
        allocate(iPiv(2,p)[p,*])
        allocate(pivotElement(p)[p,*])
#ifdef PIVOT
          if(myP==B) swap(1:n)=LU%a(k,1:n)
          if(myQ==B .and. myP>=B) then
            if(myP==B) i0=k
            if(myP>B)  i0=1
            i0 = maxLoc(abs(LU%a(i0:m,k)),1)+i0-1
            iPiv(1,myP)[B,B] = i0
            pivotElement(myP)[B,B] = abs(LU%a(i0,k))
            sync all
            if(myP==B) then
              j0 = maxLoc(pivotElement(B:p),1)+B-1
              iPiv(1,B) = iPiv(1,j0)
              iPiv(2,B) = j0
            end if
           else
            sync all
          end if

          sync all

          iPiv(1:2,myP) = iPiv(1:2,B)[B,B]

          if(myP==B .and. myQ==B) then
            iPivot(1,k)    = iPiv(1,myP)
            iPivot(2,k)    = iPiv(2,myP)
          end if

          if(iPiv(2,myP) == myP) then
            tempRow(1:n) = LU%a(iPiv(1,myP),1:n)
            LU%a(iPiv(1,myP),1:n) = swap(1:n)[B,myQ]
            swap(1:n)[B,myQ] = tempRow(1:n)
          end if

          sync all

          if(myP==B) LU%a(k,1:n) = swap(1:n)
```

```
#else
            if(myP==B .and. myQ==B) then
                iPivot(1,k) = k
                iPivot(2,k) = myP
            end if
#endif
            :
        LU%pivot = iPivot
            :
```

As shown in Listing 9.9, each image involved in the search for the pivot first finds its local pivot element and sends the value of the pivot and the local row index to the diagonal image. After synchronization, the diagonal image determines the image that holds the global pivot element and places the information in a co-array variable. After another synchronization, each image obtains the pivot information and determines if it is in the same row of images as the one holding the pivot element. If so, the image exchanges data with the diagonal image.

The function records the pivot information in the pivot array component of the result matrix. For each global row index, this array contains the local row index and image index of the pivot element. This information is used for the forward and backward solution operations.

9.7 Solving triangular systems of equations

Once a matrix has been decomposed into the product of a lower and an upper triangular matrix, solving the system of equations reduces to the solution of two triangular systems. As shown in Listing 9.10, the sequential algorithm first descends the diagonal applying forward substitution in place, with pivots as required. It then ascends the diagonal applying backward substitution.

Listing 9.10: Forward solve and backward substitution for sequential code.

```
do j=1,n
  if(ipiv(j)/=j) then
    temp = b(j)
    b(j) = b(iPiv(j))
    b(iPiv(j)) = temp
  end if
  do i=1,j-1
    b(j) = b(j) - a(j,i)*b(i)
  end do
end do
do j=n,1,-1
  do i=j+1,n
```

```
      b(j) = b(j) - a(j,i)*b(i)
   end do
   b(j) = b(j)/a(j,j)
end do
```

The parallel algorithm first works down the diagonal blocks as shown in Listing 9.11. Indeed, from (9.18) the lower-triangular part has block-partitioned form,

$$\sum_{\gamma=1}^{\alpha} L_\gamma^\alpha y^\gamma = b^\alpha , \quad \alpha = 1,\ldots,q . \tag{9.19}$$

The parallel algorithm, therefore, requires communication with images to the left as described by the equations,

$$
\begin{aligned}
L_1^1 y^1 &= b^1 \\
L_2^2 y^2 &= b^2 - L_1^2 y^1 \\
L_3^3 y^3 &= b^3 - L_1^3 y^1 - L_2^3 y^2 \\
&\vdots \\
L_q^q y^q &= b^q - L_1^q y^1 - L_2^q y^2 - L_3^q y^3 - \cdots - L_{q-1}^q y^{q-1}
\end{aligned}
\tag{9.20}
$$

Synchronization is required because lower images along the diagonal must wait for higher images to finish. At each step, the algorithm applies the pivot information to the data in the right-hand-side vector.

Listing 9.11: Solving triangular systems in parallel.

```
module DenLinAlg
  use ClassDenseMatrix
  use collectives , only : maxToAll , sumToAll , gatherToAll

  implicit none

  integer ,parameter ,private   :: wp   = kind (1.0d0)
  integer ,parameter ,private   :: zero = 0.0_wp

contains
                  :
function denseLUSolve (A,b)  result (x)
   Class (DenseMatrices) ,intent (in)  :: A
   real (wp) ,intent (in)                :: b(:)
   real (wp) ,allocatable                :: x(:)
   integer ,  allocatable                :: pivot (: ,:)
   real (wp) ,allocatable                :: tempA (: ,:) [: ,:]
   real (wp) ,allocatable                :: tempX (:) [: ,:]
   real (wp) ,parameter                  :: zero = 0.0_wp
   real (wp)                             :: temp
   integer                               :: i ,j ,k
   integer                               :: L ,m ,n
```

```
   integer                            :: p,piv,q,r
   integer                            :: myP,myQ
select type(A)
  class is (DenseMatrix)
  myP = A%myP
  myQ = A%myQ
  m   = A%localRowDim
  n   = A%localColumnDim
  r   = A%columnRemainder
  pivot = A%pivot
  allocate(tempA(A%maxDims(1),A%maxDims(2))[A%coDim1,*])
  allocate(tempX(1:A%maxDims(1))[A%coDim1,*])
  allocate(x(m))
  tempA(1:m,1:n) = A%a(1:m,1:n)
  tempX(1:m)     = b(1:m)

  sync all

  do p=1,A%coDim1
    if(myP==p .and. myQ==p) then
      tempX(1:m) = tempX(1:m)[p,1]
      do j=1,m
        k    = pivot(1,j)
        piv  = pivot(2,j)
        temp = tempX(j)
        if(piv==p) then
          tempX(j)  = tempX(k)
          tempX(k) = temp
        else
          tempX(j)          = tempX(k)[piv,1]
          tempX(k)[piv,1] = temp
        end if
      end do
      L=A%maxDims(2)
      do q=1,p-1
        tempX(1:m) = tempX(1:m) &
        - matmul(tempA(1:m,1:L)[p,q],tempX(1:L)[q,q])
        if(q==r) L=L-1
      end do
      do j=1,m
        do i=1,j-1
          tempX(j) = tempX(j) - tempA(j,i)*tempX(i)
        end do
      end do
    end if

    sync all

  end do
```

```
    do p=A%coDim2,1,-1
      if(myP==p .and. myQ==p) then
        L=A%maxDims(2)
        if(r/=0) L=L-1
        do q=A%coDim2,p+1,-1
          if(q==r) L=L+1
          tempX(1:m) = tempX(1:m) &
          - matmul(tempA(1:m,1:L)[p,q],tempX(1:L)[q,q])
        end do
        do j=m,1,-1
          do i=j+1,m
            tempX(j) = tempX(j) - tempA(j,i)*tempX(i)
          end do
          tempX(j) = tempX(j)/tempA(j,j)
        end do
      end if

      sync all

    end do
    if(A%me<=A%coDim1) then
      x(1:m) = tempX(1:m)[myP,myP]
    else
      x(1:m) = 0.0_wp
    end if
  end select
end function denseLUSolve
    :
end module DenLinAlg
```

The algorithm works back up the diagonal solving the system of equations,

$$\sum_{\beta=\gamma}^{q} U_{\beta}^{\gamma} x^{\beta} = y^{\gamma} , \quad \gamma = q,\ldots,1 , \tag{9.21}$$

applying the same kind of sychronization at each step. Images above the diagonal must wait for images below the diagonal to finish as described by the equations,

$$
\begin{aligned}
y^1 - U_2^1 x^2 - U_3^1 x^3 - \cdots - U_{q-1}^1 x^{q-1} - U_q^1 x^q &= U_1^1 x^1 \\
y^2 - U_3^2 x^3 \cdots - U_{q-1}^2 x^{q-1} - U_q^2 x^q &= U_2^2 x^2 \\
&\vdots \\
y^{q-2} - U_{q-1}^{q-2} x^{q-1} - U_q^{q-2} x^q &= U_{q-2}^{q-2} x^{q-2} \\
y^{q-1} - U_q^{q-1} x^q &= U_{q-1}^{q-1} x^{q-1} \\
y^q &= U_q^q x^q
\end{aligned}
\tag{9.22}
$$

9.8　Exercises

1. Write code for the final procedure for the dense matrix class.

2. What changes must be made to the definition of the dense matrix class to define a complex dense matrix class? With 32-bit precision?

3. Design procedures to read/write a dense matrix from/to a file.

4. Modify the matrix-matrix multiplication function shown in Listing 9.5 by adding checks that the three matrices and the image count satisfy the conformal rules derived in Section 9.4.

5. Modify the matrix-matrix multiplication function shown in Listing 9.5 such that each processor first multiplies the two blocks it owns and then cycles through the other blocks in an offset fashion.

6. How can the matrix-vector multiplication operation be implemented using the dense matrix class?

7. Implement the algorithms for matrix-matrix multiplication as listed at the end of Section 9.4. What are the restrictions on the co-dimensions in each case? Under what circumstances is one version better than the others?

8. Verify the placement of synchronization points in the LU-decomposition function shown in Listing 9.7. Can any of them be removed?

9. Verify the placement of synchronization points in the pivot code shown in Listing 9.9 and in the solver code in Listing 9.11.

Chapter 10

The Matrix Transpose Operation

The matrix-transpose operation is an important operation found in many parallel application codes. Although it is one of the simplest operations in linear algebra, it is one of the hardest operations to perform efficiently in parallel because it stresses not only local memory but also the network that connects remote memory across the machine. It often becomes a major performance bottleneck.

This chapter discusses the issues that arise when implementing the matrix-transpose operation for block-partitioned matrices. It describes how the transpose operation arises in spectral methods for solving partial differential equations. The two-dimensional fast Fourier transform algorithm is a typical example, but the issue arises as well with other algorithms. The transpose operation is often the major limit to scalability for spectral methods. The chapter also includes a performance model for the transpose operation.

10.1 The transpose operation

A block-partitioned matrix is a sum of local blocks,

$$A = \sum_{\alpha=1}^{q} \sum_{\beta=1}^{r} P_\alpha A_\beta^\alpha P^\beta , \qquad (10.1)$$

as already defined by identity (9.2). The logical image grid $[q, r]$ obeys the constraint,

$$p = q \times r , \qquad (10.2)$$

to ensure that the map between grid indices and image indices is a one-to-one map with grid indices mapped to co-dimension indices.

The transpose operation on both sides of Equation (10.1) yields the transposed matrix,

$$A^T = \sum_{\beta=1}^{r} \sum_{\alpha=1}^{q} P_\beta (A_\alpha^\beta)^T P^\alpha , \tag{10.3}$$

recalling the transpose relationship (4.17),

$$(P^\alpha)^T = P_\alpha , \tag{10.4}$$

between forward and reverse partition operators. The co-dimensions for the transposed matrix represent images decomposed into a logical $[r, q]$ grid. On the other hand, transposing the matrix before applying the partition operators yields the equation,

$$A^T = \sum_{\gamma=1}^{s} \sum_{\delta=1}^{t} P_\gamma (A^T)_\delta^\gamma P^\delta . \tag{10.5}$$

The logical grid $[s, t]$ in this case obeys the constraint,

$$p = s \times t , \tag{10.6}$$

again to maintain a one-to-one map.

For these two representations of the matrix to be the same, the transposed blocks must obey the relationship,

$$(A^T)_\delta^\gamma = (A_\beta^\alpha)^T . \tag{10.7}$$

For these blocks to be the same size, the columns of the transposed matrix must be partitioned the same way as the rows of the original matrix, and the rows of the transposed matrix the same way as the columns of the original matrix. The two image grids, therefore, must be transposed such that

$$[q, r] = [t, s] , \tag{10.8}$$

implying that $t = q$ and $s = r$. The blocks of the transposed matrix, therefore, are related to the blocks of the original matrix by the formula,

$$(A^T)_\alpha^\beta = (A_\beta^\alpha)^T , \quad \alpha = 1, \ldots, q , \quad \beta = 1, \ldots, r . \tag{10.9}$$

Figure 10.1 shows the transpose operation under these constraints.

Listing 10.1 shows code for the transpose operation. The function creates the result object my invoking the constructor function with transposed matrix size and transposed co-dimensions obtained from the incoming matrix object. The constructor function takes care of the details and allocates a local block with the correct shape and size for each image. An allocatable co-array variable holds the local transposed block. It must be allocated with the same size and shape by every image. To comply with this constraint, the function uses the maximum block dimensions calculated and stored by the constructor function.

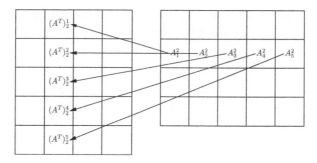

FIGURE 10.1: The matrix transpose operation for a block-partitioned matrix. The original matrix is partitioned into $20 = 4 \times 5$ blocks. The transposed matrix is partitioned into $20 = 5 \times 4$ blocks. The transposed local block from image $[\alpha, \beta]$ becomes a local block on image $[\beta, \alpha]$.

Each image transposes its own local block and places it in the temporary co-array variable. After a synchronization point, each image pulls data from its partner using its own transposed co-dimension indices. Alternatively, it could send a block to its partner who transposes it on the other side. Deallocation of the temporary co-array buffer generates a hidden synchronization point before the function returns.

Listing 10.1: The matrix transpose operation for block-partitioned matrices.

```
Type(DenseMatrix) function denseTranspose(A) result(AT)
   Class(DenseMatrix),intent(in) :: A
   real(wp),allocatable :: temp(:,:)[:,:]
   integer :: dim(2)
   integer :: m,n
     AT = DenseMatrix(A%globalColumnDim,A%globalRowDim, &
           A%coDim2,A%coDim1,A%haloWidth)
     dim = A%maxDims
     allocate(temp(dim(2),dim(1))[A%coDim1,*])
     m = A%localRowDim
     n = A%localColumnDim
     temp(1:n,1:m) = transpose(A%a(1:m,1:n))
     sync all
     m = AT%localRowDim
     n = AT%localColumnDim
     AT%a(1:m,1:n) = temp(1:m,1:n)[AT%myQ,AT%myP]
     deallocate(temp)
   end if
end function denseTranspose
```

A special case occurs when the original matrix is partitioned by rows and the transposed matrix is partitioned by columns,

$$(P^\alpha A)^T = A^T P_\alpha \,, \tag{10.10}$$

or vice versa, when the original matrix is partitioned by columns and the transposed matrix is partitioned by rows,

$$(AP_\alpha)^T = P_\alpha A^T .$$ (10.11)

Rows and columns are just swapped on the same image as shown in Figure 10.2 with no communication between images.

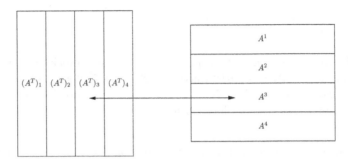

FIGURE 10.2: The matrix transpose operation for a matrix partitioned by rows on the right to a matrix partitioned by columns on the left. Each image holds the same local block transposed locally with no communication between images.

Although this form of the transpose operation is easy to understand and easy to program, it is not the typical case. In a typical application, both the matrix and its transpose are partitioned the same way either by rows or by columns. In that case, the transpose operation is a little harder to understand as discussed in the Section 10.2.

10.2 A row-partitioned matrix transposed to a row-partitioned matrix

When both the original matrix and the transposed matrix are partitioned by rows, or by columns, images must communicate with each other [26]. These cases require the use of virtual blocks within each local block. For the row-partitioned case, apply the partition operator P^α from the left,

$$A^\alpha = P^\alpha A , \quad \alpha = 1, \ldots, p ,$$ (10.12)

and assign the partition index α to the image with the same index. Application of the reverse partition operator P_β from the right,

$$A^\alpha_\beta = P^\alpha A P_\beta , \quad \beta = 1, \ldots, p ,$$ (10.13)

yields the set of local blocks. To maintain the one-to-one correspondence between partition indices and image indices, only one index α or β can correspond to an image index. Adopting the convention that a superscript index is assigned to the image index while a subscript index is assigned to a virtual block within that image resolves this problem of notation. The transpose operation, then, switches the partition indices,

$$(A_\beta^\alpha)^T = (P^\alpha A P_\beta)^T = P^\beta A^T P_\alpha = (A^T)_\alpha^\beta \,, \tag{10.14}$$

and assigns the partition of the transposed matrix labeled by index β, now as a superscript, to image β with virtual block within that image labeled by index α, now as a subscript.

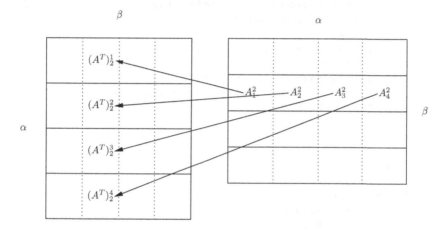

FIGURE 10.3: Matrix transpose with both matrices partitioned by rows. Communication among images is required since the subpartitions owned by each image must be transposed and, except for one block on each image, must be moved to another image. The direction of the arrows implies that each image sends each of its virtual blocks to the appropriate partner.

Figure 10.3 shows a picture of the transpose operation between two row-partitioned matrices. It is an example of the all-to-all operation, every image exchanges blocks of data with every other image. Listing 10.2 shows code for the transpose operation. It is similar to the previous example. The function invokes the constructor function to create the transposed result object using information from the object that invoked the function. Each image allocates a temporary co-array variable large enough to hold the local transposed block on each image. Each image transposes its local data into this temporary co-array variable and then pulls the transposed data into the array component of the result object. Notice carefully the use of the offset variables to reference virtual blocks. Each image contains one block of data that just needs to be transposed locally without incurring traffic across the network.

Listing 10.2: Transposing a row-partitioned matrix to a row-partitioned matrix.

```
Type(DenseMatrix) function denseTranspose(A) result(AT)
  Class(DenseMatrix),intent(in) :: A
  real(wp),allocatable :: temp(:,:)[:,:]
  real(wp),allocatable :: tempA(:,:)[:]
  integer :: dim(2)
  integer :: m,n
  integer :: j,K_0,L_0,q,r

  AT = denseMatrix(A%globalColumnDim,A%globalRowDim, &
      coDim1=A%coDim1,coDim2=A%coDim2,              &
      haloWidth=A%haloWidth)
  m = A%localRowDim
  n = A%localColumnDim
  dim = A%maxDims
  allocate(tempA(dim(2),dim(1))[*])
  tempA(1:n,1:m) = transpose(A%a(1:m,1:n))
  m = AT%localRowDim
  r = A%RowRemainder
  n = dim(1)

  sync all

  do q=1,AT%p
      K_0 = (q-1)*n
      if(q>r) K_0=K_0+r
      if(q == AT%me) then
        AT%a(1:m,K_0+1:K_0+n) = &
            tempA(AT%K_0+1:AT%K_0+m,1:n)
      else
        AT%a(1:m,K_0+1:K_0+n) = &
            tempA(AT%K_0+1:AT%K_0+m,1:n)[q]
      end if
      if(q==r) n=n-1
  end do
  deallocate(tempA)
end function denseTranspose
```

The function could transpose data either before or after moving it across the network. Or perhaps it could transpose the data as it moves across the network. Since most current networks perform very badly for non-contiguous transfers, the last option is probably a bad idea even though it is simple to write the operation using co-array syntax.

10.3   The Fast Fourier Transform

The matrix-transpose operation is a fundamental component of spectral methods for solving partial differential equations defined over multi-dimensional domains [40] [100]. The nice property of spectral transforms is that they convert differential operations into algebraic operations. But they must be used carefully. Calling a function from a black box library can be dangerous if the programmer does not understand the conventions used by a particular library. In most libraries, phases are restricted to their principal values in the complex plane $(-\pi, \pi]$, but normalization conventions may be different for different libraries.

The Fourier transform is a common method employed, but the issues involved apply just as well to other transforms. It can be described as matrix multiplication, but the Fourier transform can be performed more efficiently using the Fast Fourier Transform (FFT) algorithm [94][95]. Libraries are publicly available for this operation [38].

Library functions perform the FFT transform for arrays of data that reside in local memory on a single processor. Each image, therefore, can apply the transform directly to the columns of a column-partitioned matrix independently of the other images with no communication between images. Transforming the rows of the matrix, however, requires a transpose operation such that the rows are contiguous in local memory. Each image applies the FFT to the columns of the transposed matrix and follows it with a second transpose operation to return the data to its original storage layout.

Listing 10.3: A two-dimensional Fast Fourier Transform.

```
use ComplexMatrix
use FFTLibrary
Type(ComplexMatrix)              :: F
Type(ComplexMatrix), allocatable :: Fhat
 p = num_images()
         :
 F = ComplexMatrix(n,n,coDim1=1,coDim2=p,haloWidth=0)
         :
       ! ... calculations in physical space ...!
         :
 Fhat = F%FFT(+1)      !  forward transform
 Fhat = Fhat%transpose()
 Fhat = Fhat%FFT(+1)
         :
       ! ... calculations in spectral space ...!
         :
 F     = Fhat%FFT(-1) !  inverse transform
 F     = F%transpose()
 F     = F%FFT(-1)
```

```
         :
!  ...  calculations  in  physical  space  ...!
         :
```

Listing 10.3 shows the skeleton of a program that performs a two-dimensional FFT. It uses a complex blocked matrix class, an extension of the abstract class for dense matrices with complex data components rather than real components. The program uses a module containing FFT libraries, and the complex object has a procedure pointer associated with one of the FFT functions in that library.

Spectral transforms in three or higher dimensions require similar partitioning and transposition strategies. More than one option is available. The programmer might want first to partition the problem along the third dimension to perform the transforms on the first two dimensions within each partition followed by a transpose operation that partitions the problem along either the first or second dimension to perform the transform along the third dimension. Another option is to cut the three-dimensional problem into pencils along one dimension by partitioning the other two dimensions into blocks and performing the transform for all data in the pencil. The problem must then be transposed into pencils along another dimension, the transpose performed for those pencils, and finally transposed into pencils along the last dimension.

10.4   Performance analysis

Measurements of execution time for the matrix-transpose operation, plotted in Figures 10.4 and 10.5 as the logarithm of the normalized execution time,

$$\tau(n,p) = t(n,p)/t(n,1) \ , \tag{10.15}$$

versus the logarithm of the image count, suggest that the normalized execution time is an exponential function of image count.

The methods of dimensional analysis provide one way to understand this exponential behavior [83]. A reasonable starting point assumes that the execution time can be described by some relationship among four quantities,

$$t(n,p) = f(w,p,b_0,b) \ . \tag{10.16}$$

The first quantity is a function of matrix size,

$$w(n) = n^2 \ell_0 \ , \tag{10.17}$$

equal to the total amount of data transposed where n is the matrix size and ℓ_0 is the length of one data element. The matrix size may or may not be a function of the image count depending on whether the case considered involves strong scaling or weak scaling. The second quantity p is the image count, a dimensionless quantity. The other two quantities are bandwidths measured in units of length per time, usually bytes per second or words per second. The first quantity b_0 represents local bandwidth within a node and the second quantity b represents network bandwidth between nodes.

Only the two dimensions of length and time enter into the relationship. Dimensional analysis says that two of the four parameters can be eliminated by picking two scale factors, one for length α_L and one the time α_T, and scaling the quantities in relationship (10.16) according to their dimensions to obtain an equivalent dimensionless relationship,

$$\alpha_T t(n, p) = f(\alpha_L w, p, \alpha_L \alpha_T^{-1} b_0, \alpha_L \alpha_T^{-1} b) \ . \tag{10.18}$$

Bandwidth has dimension of length divided by time so it is multiplied by the scale factors $\alpha_L \alpha_T^{-1}$. The time is only multiplied by the scale factor α_T, and the unit of length is multiplied by the single scale factor α_L. The problem size n and the image count p are dimensionless quantities and are not scaled.

Picking the scale factors such that

$$\alpha_L w = 1 \ , \quad \alpha_L \alpha_T^{-1} b_0 = 1 \ , \tag{10.19}$$

eliminates the first and third quantities from the relationship. Solutions for the scale factors from these two equations yield the results,

$$\alpha_L = 1/w \ , \quad \alpha_T = b_0/w \ , \tag{10.20}$$

with dimensions of inverse length for the first scale factor and inverse time for the second.

The scale factor for time is, in fact, the reciprocal of the execution time on one image,

$$t(n, 1) = w(n)/b_0 \ . \tag{10.21}$$

The fourth scaled quantity has the value,

$$\beta = b/b_0 \ , \tag{10.22}$$

the ratio of remote bandwidth to local bandwidth. The scaled relationship, then, has the value,

$$\tau(n, p) = f(1, p, 1, \beta) \ , \tag{10.23}$$

a dimensionless function of just two dimensionless quantities, p and β. The observed exponential behavior, therefore, might be represented by the function,

$$\tau(n, p) = b(\beta, p) p^{a(\beta, p)} \ , \tag{10.24}$$

with parameters $b(\beta, p)$ and $a(\beta, p)$ as functions of the two parameters β and p. How well this function represents the execution time as a function of processor count is determined by how well it matches actual measured values.

10.5   Strong scaling

For the strong-scaling case, the problem size is fixed as the image count changes. It is reasonable to expect the normalized execution time to decrease with image count p because the local block size decreases. And it should increase as the ratio of remote bandwidth to local bandwidth β decreases.

Indeed, Figure 10.4 shows such behavior for measured times with fixed matrix size $n = 2500$. The dotted lines through the measured values correspond to the function,

$$\tau(n,p) = \begin{cases} p^{-5/6} & p < 32 \\ 4p^{-4/6} & p \geq 32 \end{cases} \qquad (10.25)$$

corresponding to Equation (10.24) with $a(\beta, p) = -5/6$, $b(\beta, p) = 1.0$ for $p < 32$ and $a(\beta, p) = -4/6$, $b(\beta, p) = 4$ for $p \geq 32$.

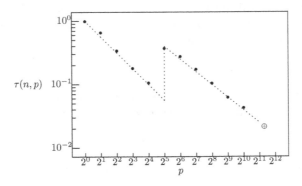

FIGURE 10.4: Normalized execution times for the matrix-transpose operation as a function of the number of images for the strong-scaling case. The matrix is square with fixed size, $n = 2500$. Bullets ($\bullet$) mark the measured values. Dotted lines through the bullets correspond to the function (10.25). The cross ($\oplus$) marks the limit of scalability at $p = n$.

This function captures the behavior of the normalized execution time quite well. A reasonable interpretation of these results is that the parameter $b(\beta, p) = 1/\beta(p)$ is the reciprocal of the relative bandwidths that increases by a factor of four outside a node at $p = 32$. Likewise, the parameter $a(\beta, p)$ is a function of the image count alone $a(p)$ and measures the contention among images using shared resources. The exponential decay rate is slightly lower outside the node than inside the node as reflected in the smaller value of the parameter.

Since the matrix is partitioned by rows, the largest number of images that can be used is $p = n$. At that point, the partition size assigned to each image is one and the addition of more images results in no decrease in execution time. The cross in Figure 10.4 marks this limit.

10.6 Weak scaling

For the weak-scaling case, the problem size changes as the image count changes. If it changes proportional to the image count,

$$n(p) = pn(1) , \qquad (10.26)$$

for some value $n(1)$ on a single image, the global matrix size increases while the number of rows assigned to each image remains fixed. The normalized execution time (10.15) is the function,

$$\tau(n(p), p) = t(n(p), p)/t(n(p), 1) , \qquad (10.27)$$

with the fixed problem size n replaced by the function $n(p)$.

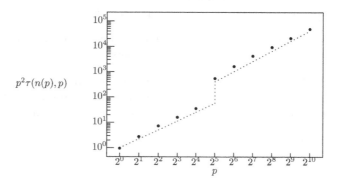

FIGURE 10.5: Execution time for the matrix-transpose operation under the fixed-partition-size constraint. Bullets ($\bullet$) mark the measured values starting from the initial matrix size $n(1) = 250$. The dotted lines correspond to function (10.30).

The execution time on a single image for a matrix of size $n(p)$ increases like the square of the number of images compared with the time for the original problem,

$$t(n(p), 1) = p^2 t(n(1), 1) . \qquad (10.28)$$

for a matrix of size $n(1)$. The normalized execution time, then, obeys the formula,

$$p^2 \tau(n(p), p) = t(n(p), p)/t(n(1), 1) , \qquad (10.29)$$

avoiding the need to measure the increasingly large execution time on a single image as the problem size increases. If the function (10.25) still holds, the measured execution time should behave like the function,

$$p^2 \tau(n(p), p) = \begin{cases} p^{2-5/6} & p < 32 \\ 4p^{2-4/6} & p \geq 32 \end{cases} \qquad (10.30)$$

Figure 10.5 shows measured execution times marked by bullets (•) with dotted lines representing function (10.30). It is somewhat surprising that the same values for the parameters represent both strong and weak scaling. This result suggests that they reflect properties of the hardware independent of the way the matrix size is scaled. Comparison of different machines, therefore, can be accomplished with modest strong-scaling experiments rather than with expensive weak-scaling experiments that require high levels of programming effort and computer time.

10.7   Exercises

1. Change the code in Listing 10.2 so that each image sends its local block to its partner with the partner transposing the block.

2. Change the code in Listing 10.2 so that the matrix blocks are transposed as they cross the network between partners.

3. Convert the dense matrix class to a complex matrix class. Include a procedure pointer for the FFT operation with an interface that allows it to be associated with a particular library version of the FFT.

4. Verify formula (10.30).

Chapter 11

The Halo Exchange Operation

The halo-exchange operation occurs frequently in parallel applications, for example, in finite difference codes and in finite element and finite volume codes. It also occurs in molecular dynamics codes where particles interact with remote particles through forces that move them from one physical domain to another.

For such problems, the programmer typically partitions physical space into domains and assigns a local domain to the image with image index equal to the partition index. Halo cells surround each local domain to hold local copies of data owned by other images. The exchange operation updates the values in these halo cells to keep them current as an algorithm advances in time.

11.1 Finite difference methods

The heat equation provides a good example of the finite difference method [72, pp. 7-22]. It is a time-dependent, second-order partial differential equation,

$$\frac{\partial u}{\partial t} = \Delta u \; , \tag{11.1}$$

for the function $u(x, t)$ defined on some physical domain, a subset of d-dimensional Euclidean space,

$$x \in \mathcal{D} \subset \mathcal{R}^d \; , \tag{11.2}$$

with, for example, Dirichlet boundary conditions,

$$u(x, t) = g(x) \; , \quad x \in \partial \mathcal{D} \; , \tag{11.3}$$

on the boundary of the domain. The initial-value problem starts with some initial function,

$$u_0(x) = u(x, t_0) \; , \tag{11.4}$$

and advances the solution as time increases.

The finite difference method approximates the time derivative with, for example, a forward difference formula,

$$\frac{\partial u}{\partial t}(x, t_{k+1}) \approx \frac{u_{k+1}(x) - u_k(x)}{\Delta t} , \tag{11.5}$$

where

$$u_k(x) = u(x, t_k) \tag{11.6}$$

for discrete values of the time,

$$t_k = t_0 + k\Delta t , \quad k = 0, 1, \dots , \tag{11.7}$$

for some time step Δt.

For the one-dimensional case, the finite element method approximates the Laplacian with, for example, a three-point centered difference formula,

$$\Delta u(x_i, t_k) \approx \frac{u_k(x_{i+1}) - 2u_k(x_i) + u_k(x_{i-1})}{(\Delta x)^2} , \quad i = 1, \dots, n , \tag{11.8}$$

at each time t_k with the space variable evaluated at the discrete points,

$$x_i = x_0 + i\Delta x , \quad i = 0, \dots, n+1 , \tag{11.9}$$

with spacing between points,

$$\Delta x = (x_{n+1} - x_0)/(n+1) . \tag{11.10}$$

Starting with the initial function, the solution advances in time according to the finite difference formula,

$$\begin{aligned} u_{k+1}(x_i) &= u_k(x_i) \\ &+ \mu \left[u_k(x_{i+1}) - 2u_k(x_i) + u_k(x_{i-1}) \right] , \quad i = 1, \dots, n , \end{aligned} \tag{11.11}$$

where

$$\mu = \Delta t/(\Delta x)^2 . \tag{11.12}$$

The time step Δt must satisfy an appropriate stability criterion given the distance Δx between space points [72, p. 17].

This method yields a solution at the interior points of the domain subject to the conditions,

$$u(x_0, t_k) = g(x_0) , \tag{11.13}$$

$$u(x_{n+1}, t_k) = g(x_{n+1}) , \tag{11.14}$$

at the left and right boundaries at each time step. The two extra points, x_0 and x_{n+1}, are halo cells that surround the solution points inside the domain. They are required both to enforce the boundary conditions and to evaluate the finite difference formula (11.11).

Some problems require periodic boundary conditions. In this case, the periodic condition,

$$u(x_1, t_k) = u(x_n, t_k) \ , \tag{11.15}$$

replaces the boundary conditions (11.13)-(11.14). The halo cells hold the values,

$$
\begin{aligned}
u(x_0, t_k) &= u(x_{n-1}, t_k) \ , & (11.16) \\
u(x_{n+1}, t_k) &= u(x_2, t_k) \ . & (11.17)
\end{aligned}
$$

The two-dimensional case is more common and more interesting. A discrete rectangular grid,

$$
\begin{aligned}
x_i &= x_0 + i\Delta x \ , & i = 0, \dots, n+1 \ , & \qquad (11.18) \\
y_j &= y_0 + j\Delta y \ , & j = 0, \dots, m+1 \ , & \qquad (11.19)
\end{aligned}
$$

with spacing between points,

$$
\begin{aligned}
\Delta x &= (x_{n+1} - x_0)/(n+1) \ , & (11.20) \\
\Delta y &= (y_{m+1} - y_0)/(m+1) \ , & (11.21)
\end{aligned}
$$

replacing the one-dimensional discrete line (11.9). With the same explicit time scheme (11.5) to advance the solution from the initial value and with a five-point centered difference formula to approximate the Laplacian, the two-dimensional finite difference formula assumes the form,

$$
\begin{aligned}
u_{k+1}(x_i, y_j) &= u_k(x_i, y_j) \\
&+ \mu_x \left[u_k(x_{i+1}, y_j) - 2u_k(x_i, y_j) + u_k(x_{i-1}, y_j) \right] \\
&+ \mu_y \left[u_k(x_i, y_{j+1}) - 2u_k(x_i, y_j) + u_k(x_i, y_{j-1}) \right] \ , \\
&\quad i = 1, \dots, n \ , \quad j = 1, \dots, m \ ,
\end{aligned}
\tag{11.22}
$$

where

$$
\begin{aligned}
\mu_x &= \Delta t/(\Delta x)^2 \ , & (11.23) \\
\mu_y &= \Delta t/(\Delta y)^2 \ , & (11.24)
\end{aligned}
$$

with appropriate conditions on the time step.

Halo cells surround the rectangle with two vertical lines on the left and right sides of the rectangular and two horizontal lines at the top and bottom. The values held in the halo cells enforce the boundary conditions,

$$
\begin{aligned}
u_k(x_0, y) &= g_0(y) \ , & (11.25) \\
u_k(x_{n+1}, y) &= g_{n+1}(y) \ , & (11.26) \\
u_k(x, y_0) &= h_0(x) \ , & (11.27) \\
u_k(x, y_{m+1}) &= h_{m+1}(x) \ , & (11.28)
\end{aligned}
$$

with four functions defined on the boundaries. The values of these functions on the boundary also enter into the evaluation of the finite difference formula (11.22).

Halo cells with depth two are needed to support a five-point difference scheme for the one-dimensional case or a nine-point scheme for the two-dimensional case. For the less common three-dimensional case, the halo cells become planes surrounding the domains.

11.2   Partitioned finite difference methods

In the one-dimensional finite difference scheme, the programmer partitions the interior points of the physical domain,

$$x = \begin{bmatrix} x_1 \\ \vdots \\ x_n \end{bmatrix} \tag{11.29}$$

by applying partition operators in the usual way,

$$x^\alpha = P^\alpha x , \tag{11.30}$$

and assigns each partition index to the image with the same image index. Figure 11.1 shows an example with halo cells appended to the ends of each local partition.

The finite difference problem for each image looks just like the original problem. At each time step, however, images require values from neighbors to the left and to the right. The halo-exchange operation updates these values with synchronization to ensure that they are current values at each time step.

The halo cells hold the values,

$$x_0^1 = g(x_0) \quad , \quad x_4^1 = x_1^2 , \tag{11.31}$$
$$x_0^2 = x_3^1 \quad , \quad x_4^2 = x_1^3 , \tag{11.32}$$
$$x_0^3 = x_3^2 \quad , \quad x_3^3 = x_1^4 , \tag{11.33}$$
$$x_0^4 = x_2^3 \quad , \quad x_3^4 = g(x_1) . \tag{11.34}$$

For periodic boundary conditions, the values on the global domain boundary are the same,

$$x_1^1 = x_2^4 , \tag{11.35}$$

and the left and right halos hold the values,

$$x_0^1 = x_1^4 , \quad x_3^4 = x_2^1 . \tag{11.36}$$

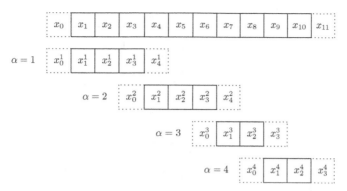

FIGURE 11.1: A partitioned one-dimensional domain with added halo cells. For the case $n = 10$, $p = 4$, the first two images own $\lceil 10/4 \rceil = 3$ points while the last two images own $\lfloor 10/4 \rfloor = 2$ points. Halo cells at the left and right ends of each local partition hold values from neighboring images. The left halo on image one holds the left boundary value and the right halo on the last image holds the right boundary value.

Figure 11.2 shows the two-dimensional case with physical space partitioned into blocks. Each block is labeled by two block indices that map to two co-dimension indices. There is a one-to-one correspondence between these indices, but the programmer must adopt a convention for how the co-dimension indices are related to the block indices. The arrows in the lower left corner of Figure 11.2 show one possible convention. If the local block corresponds to co-dimension indices [myP,myQ], increasing the first co-dimension index, [myP+1,myQ], corresponds to the up direction while increasing the second co-dimension index, [myP,myQ+1], corresponds to the right direction. The programmer, not the co-array model, sets the convention of what is up and what is down, what is left and what is right, and should document it to remove any confusion in the code.

The programmer must handle data in the corners carefully. They correspond to partners one over and one up or one over and one down. These points can be exchanged with four separate transfers around the corners of each block using co-dimension indices such as [myP+1,myQ+1] to reference data from the partner to the northeast. The relationship between halo cells shown in Figure 11.2, however, suggests a more elegant way to fill in the corners of the halos. Once the interior points have been copied into the halo cells, say from neighbors up and down, the values that need to be copied into the corners are in the upper and lower halos of the neighbors left and right. The second halo exchange, therefore, copies not only the interior points into the halo cells but also the full halo cell from neighbors to the left and right.

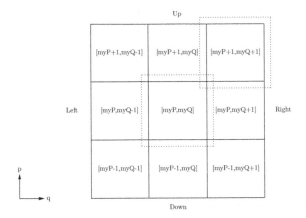

FIGURE 11.2: Representation of a two-dimensional domain decomposition using co-dimensions for up-down-left-right directions. The arrows in the lower left corner indicate the convention for increasing co-dimension indices.

11.3 The halo-exchange subroutine

Listing 11.1 shows code for the halo-exchange operation. It corresponds to the procedure pointer included in the dense matrix class shown in Listing 9.2 and satisfies the interface block defined there. A block matrix object represents the points of the two-dimensional physical domain. The constructor allocates local blocks with the correct dimensions with specified halo widths and sets the procedure pointer to the exchange subroutine. The exchange subroutine allocates two co-array buffers to communicate halo arrays, first in the up-down direction and then in the left-right direction. Images with blocks on the boundary of the domain leave the boundary values unchanged. Allocation of the co-array variables provides a synchronization point before any exchange takes place. Deallocation followed by allocation between the two exchange directions provides another synchronization point, and deallocation at the end of the procedure implies synchronization before control returns to the caller.

Listing 11.1: The halo-exchange subroutine for block matrix objects.

```
subroutine denseExchange(A)
  Class(DenseMatrix),intent(inout) :: A
  integer :: myP,myQ,m,n,w,p,q
  real(wp),allocatable :: tempA(:,:)[:,:]
  real(wp),allocatable :: tempB(:,:)[:,:]
  integer :: dims(2)
    myP  = A%myP
    myQ  = A%myQ
    w    = A%haloWidth
```

```
m     = A%localRowDim
n     = A%localColumnDim
p     = A%coDim1
q     = A%coDim2
dims  = A%maxDims+2*w

allocate(tempA(w,dims(2))[p,*])
allocate(tempB(w,dims(2))[p,*])
tempA(1:w,1:n)  =  A%a(m-w+1:m,1:n)
tempB(1:w,1:n)  =  A%a(1:w,1:n)
sync all
if(myP<A%coDim1)  A%a(1-w:0,1:n)    = &
                  tempA(1:w,1:n)[myP+1,myQ]
if(myP>1)         A%a(m+1:m+w,1:n) = &
                  tempB(1:w,1:n)[myP-1,myQ]
deallocate(tempA)
deallocate(tempB)
allocate(tempA(1-w:dims(1)+w,1:w)[A%coDim1,*])
allocate(tempB(1-w:dims(1)+w,1:w)[A%coDim1,*])
tempA(1-w:m+w,1:w)  =  A%a(1-w:m+w,n-w+1:n)
tempB(1-w:m+w,1:w)  =  A%a(1-w:m+w,1:w)
sync all
if(myQ > 1)        A%a(1-w:m+w,1-w:0)   = &
                   tempA(1-w:m+w,1:w)[myP,myQ-1]
if(myQ < A%coDim2) A%a(1-w:m+w,n+1:n+w) = &
                   tempB(1-w:m+w,1:w)[myP,myQ+1]
deallocate(tempA)
deallocate(tempB)
end subroutine denseExchange
```

Solving the heat equation, then, is straightforward as shown by the code sample in Listing 11.2. It is a direct transcription of the finite element formula (11.22), using the local block dimensions rather than the global dimensions, with the insertion of the halo-exchange at each time step.

Listing 11.2: Solving the two-dimensional heat equation.

```
use ClassDenseMatrix
Type(DenseMatrix) :: U
    :
    U = DenseMatrix(m,n,coDim1=p,coDim2=q,haloWidth=w)
    :
    do while(t < tMax)
        call U%exchange()
        U%a(1:mLocal,1:nLocal)    = U%a(:,:)        &
          + muX*(U%a(2:mLocal+w,:)  - 2*U%a(:,:)     &
                           + U%a(1-w:mlocal,:))      &
          + muY*(U%a(:,2:nLocal+w)  - 2*U%a(:,:)     &
                           + U%a(:,1-w:nlocal))
        t = t + deltaT
```

```
        end do
```

11.4   Exercises

1. Implement a parallel algorithm for integrating the one-dimensional heat equation using a five-point finite difference formula to approximate the Laplace operator.

2. Use a nine-point formula for the two-dimensional equation.

3. Solve the heat equation on the surface of a cylinder, that is, impose periodic boundary conditions in the x-direction and two-point boundary conditions in the y-direction.

4. Impose periodic boundary conditions in both the x-direction and the y-direction and solve the heat equation on the surface of a torus.

5. What changes would you need to make to the code shown in Listing 11.1 to exchange halo cells first in the left-right direction followed by the up-down direction?

Chapter 12

Subpartition Operators

The co-array programming model is based on a correspondence between a set of partition indices labeling blocks of a decomposed data structure and a set of image indices labeling replications of a program across physical processors. If the two index sets have the same size, each image owns a unique block, and co-dimensions provide a convenient representation of the one-to-one map between the two sets.

Requiring a one-to-one map between index sets, however, is too restrictive. The conformal rules for matrix-matrix multiplication, for example, imply that the number of images must be square and that each of the two co-dimensions for co-array variables must equal the square root of the number of images.

Subpartition operators remove the one-to-one constraint by mapping the set of global indices to the set of image indices in a two-step process. The first step partitions the global index set into blocks independent of the number of images. The second step assigns the block indices to image indices. The conformal rules for linear algebra, then, require the rows and columns of the matrices to be blocked correctly without placing restrictions on co-dimensions or on the number of images. Algorithms become more complicated, but the added flexibility is necessary to obtain good performance for complicated applications.

12.1 Subpartition operators

The first step in the process partitions a global index set,

$$N = \{1, \ldots, n\} , \tag{12.1}$$

into q blocks,

$$N^b = Q^b N \ , \quad b = 1, \ldots, q \ , \tag{12.2}$$

independent of the number of images. These operators define a partition of unity when summed over blocks,

$$\sum_{b=1}^{q} Q_b Q^b = I_n \ , \tag{12.3}$$

where I_n is the identity operator of size n.

Operator Q^b extracts a set of local block indices from the set of global indices, according to rules that depend on the remainder,

$$r = n \bmod q \ , \tag{12.4}$$

and on the ceiling function,

$$m_0 = \lceil n/q \rceil \ . \tag{12.5}$$

Indeed, substitution of the block index b for the image index α in formula (3.20) yields a map from a global index k to a specific block index $b(k)$ by the formula,

$$b(k) = \begin{cases} \lfloor \frac{k-1}{m_0} \rfloor + 1 & r = 0 \\ \lfloor \frac{k-1}{m_0} \rfloor + 1 & r > 0 \quad k \le m_0 r \\ \lfloor \frac{k-r-1}{m_0-1} \rfloor + 1 & r > 0 \quad k > m_0 r \end{cases} \ , \quad k = 1, \ldots, n \ . \tag{12.6}$$

With the same substitution, formula (3.19) yields a formula for the size of each block,

$$m(b) = \begin{cases} m_0 & r = 0 \quad b = 1, \ldots, q \\ m_0 & r > 0 \quad b = 1, \ldots, r \\ m_0 - 1 & r > 0 \quad b = r+1, \ldots, q \end{cases} \ , \tag{12.7}$$

and formula (3.21) yields a formula for the global base index,

$$k_0(b) = \begin{cases} (b-1)m(b) & r = 0 \quad b = 1, \ldots, q \\ (b-1)m(b) & r > 0 \quad b = 1, \ldots, r \\ (b-1)m(b) + r & r > 0 \quad b = r+1, \ldots, q \end{cases} \ , \tag{12.8}$$

for each block. Local block indices, then, correspond to global indices,

$$N^b = \{ k_0(b) + 1, \ldots, k_0(b) + m(b) \} \ , \quad b = 1, \ldots, q \ , \tag{12.9}$$

according to the formula,

$$i(k) = k - k_0(b(k)) \ , \quad k = 1, \ldots, n \ . \tag{12.10}$$

The inverse map from a local index back to the corresponding global index obeys the formula,

$$k(i) = k_0(b) + i \ , \quad i = 1, \ldots, m(b) \ , \quad b = 1, \ldots, q \ . \tag{12.11}$$

12.2   Assigning blocks to images

The second step in the process partitions the block indices,

$$B = \{1, \ldots, q\} \,, \tag{12.12}$$

and assigns the blocks to image indices,

$$B^\alpha = P^\alpha B \,, \quad \alpha = 1, \ldots, p \,, \tag{12.13}$$

using a set of partition operators that satisfies a partition of unity,

$$\sum_{\alpha=1}^{p} P_\alpha P^\alpha = I_q \,, \tag{12.14}$$

where I_q is the identity operator of size q.

The rules for the assignment of block indices to image indices depend on the remainder,

$$s = q \bmod p \,, \tag{12.15}$$

and on the ceiling function,

$$\ell_0 = \lceil q/p \rceil \,. \tag{12.16}$$

A block index maps to an image index according to the formula,

$$\alpha(b) = \begin{cases} \lfloor \frac{b-1}{\ell_0} \rfloor + 1 & s = 0 \\[2mm] \lfloor \frac{b-1}{\ell_0} \rfloor + 1 & s > 0 \quad b \le \ell_0 s \\[2mm] \lfloor \frac{b-s-1}{\ell_0-1} \rfloor + 1 & s > 0 \quad b > \ell_0 s \end{cases} \,, \quad b = 1, \ldots, q \,. \tag{12.17}$$

The number of blocks assigned to image α has the value,

$$\ell(\alpha) = \begin{cases} \ell_0 & s = 0 \quad \alpha = 1, \ldots, p \\ \ell_0 & s > 0 \quad \alpha = 1, \ldots, s \\ \ell_0 - 1 & s > 0 \quad \alpha = s+1, \ldots, p \end{cases} \,, \tag{12.18}$$

with base block index,

$$b_0(\alpha) = \begin{cases} (\alpha-1)\ell(\alpha) & s = 0 \quad \alpha = 1, \ldots, p \\ (\alpha-1)\ell(\alpha) & s > 0 \quad \alpha = 1, \ldots, s \\ (\alpha-1)\ell(\alpha) + s & s > 0 \quad \alpha = s+1, \ldots, p \end{cases} \,. \tag{12.19}$$

Image α, therefore, owns $\ell(\alpha)$ blocks,

$$B^\alpha = \{N^{b_0(\alpha)+1}, \ldots, N^{b_0(\alpha)+\ell(\alpha)}\} \,, \quad \alpha = 1, \ldots, p \,, \tag{12.20}$$

with local block indices,

$$j(b) = b - b_0(\alpha(b)) \, , \quad b = 1, \ldots, q \, . \tag{12.21}$$

The inverse map from local block index back to the original block index obeys the formula,

$$b(j) = b_0(\alpha) + j \, , \quad j = 1, \ldots, \ell(\alpha) \, , \quad \alpha = 1, \ldots, p \, . \tag{12.22}$$

12.3   Combined effect of the two partition operations

The combined formulas from the two partition operations yield a forward map taking a global index to a local index within a block assigned to an image. Indeed, Equation (12.6) maps global index k to a block $b(k)$. Equation (12.17) assigns block $b(k)$ to image $\alpha(b(k))$ with local block index $j(b(k))$ from Equation (12.21) and local element index $i(k)$ within the block from Equation (12.10).

In the opposite direction, an image maps a local index within one of its blocks back to the original global index by reversing these formulas. Indeed, Equation (12.22) maps local block index j to global block index $b(j)$ and Equation (12.11) maps a local index i within a block back to the global index k by the formula,

$$k = k_0(b_0(\alpha) + j) + i \, , \quad i = 1, \ldots, m(b_0(\alpha) + j) \, . \tag{12.23}$$

Image α, in its local block j, owns global indices,

$$\begin{aligned} B_j^\alpha = \{ k_0(b_0(\alpha) + j) \quad &+ \quad 1, \ldots, k_0(b_0(\alpha) + j) + m(b_0(\alpha) + j) \} \, , \\ \alpha \quad &= \quad 1, \ldots, p \, , \quad j = 1, \ldots, \ell(\alpha) \, . \end{aligned} \tag{12.24}$$

12.4   Permuted distributions

It is sometimes necessary to permute the block indices before assigning them to images. Before permutation, each global index k maps to a block $b(k)$ according to Equation (12.6) without change. Equation (12.7) defines the block size $m(b)$, and Equation (12.8) defines the base index $k_0(b)$.

A permutation operator shuffles each block index to a new value,

$$c = \pi(b) \, , \quad b = 1, \ldots, q \, . \tag{12.25}$$

The local index within the permuted block has the value,

$$i(k) = k - k_0(\pi^{-1}(c)) , \quad k = 1, \ldots, n \tag{12.26}$$

according to Equation (12.10) where the inverse permutation operation, defined by the formula,

$$\pi^{-1}(\pi(c)) = c , \tag{12.27}$$

maps the permuted block index back to the original block index. The inverse map from a local index within the permuted block back to the global index has the value,

$$k(i) = k_0(\pi^{-1}(c)) + i , \quad i = 1, \ldots, m(\pi^{-1}(c)) , \quad c = 1, \ldots, q , \tag{12.28}$$

according to formula (12.11).

Equation (12.17) assigns the permuted blocks to images with the number of blocks $\ell(\alpha)$ assigned to image α defined by Equation (12.18) and base block index $b_0(\alpha)$ defined by Equation (12.19). Equation (12.21) defines the local block indices on image α,

$$j(c) = c - b_0(\alpha(c)) , \quad c = 1, \ldots, q , \tag{12.29}$$

and Equation (12.22) defines the inverse map,

$$b(j) = \pi^{-1}(b_0(\alpha) + j) , \quad j = 1, \ldots, \ell(\alpha) , \quad \alpha = 1, \ldots, p . \tag{12.30}$$

Image α owns a set of blocks,

$$C^\alpha = \{N^{\pi^{-1}(b_0(\alpha)+1)}, \ldots, N^{\pi^{-1}(b_0(\alpha)+\ell(\alpha))}\} , \alpha = 1, \ldots, p \tag{12.31}$$

with global indices in each block,

$$k = k_0(\pi^{-1}(b_0(\alpha) + j)) + i , \quad i = 1, \ldots, m(\pi^{-1}(b_0(\alpha) + j)) , \tag{12.32}$$

labeled with local indices,

$$\begin{aligned} C_j^\alpha &= \{k_0(\pi^{-1}(b_0(\alpha) + j)) + 1, \\ &\quad \ldots , k_0(\pi^{-1}(b_0(\alpha) + j)) + m(\pi^{-1}(b_0(\alpha) + j))\} , \\ \alpha &= 1, \ldots, p , \quad j = 1, \ldots, \ell(\alpha) . \end{aligned} \tag{12.33}$$

The permutation operator and the partition operator together determine the distribution of blocks across images. Insertion of a permutation operation into the formulas does not destroy the partition of unity. Indeed, the matrix representing the permutation operation is an orthogonal matrix,

$$\Pi^T \Pi = I_q , \tag{12.34}$$

with each row containing a single nonzero entry equal to one at the permuted block index corresponding to that row. Insertion of this identity operator into the partition of unity,

$$\sum_q (Q_q \Pi^T)(\Pi Q^q) = I_q \ , \qquad (12.35)$$

maintains the identity with the permuted operator,

$$R^q = \Pi Q^q \ , \qquad (12.36)$$

replacing the unpermuted operator.

12.5 The cyclic distribution

The cyclic distribution results from a permutation operator defined implicitly by assigning blocks to images as if dealt from a deck of cards. For p images, block b is assigned to an image according to the formula,

$$\alpha(b) = 1 + (b-1) \bmod p \ , \quad b = 1, \ldots, q \ , \qquad (12.37)$$

as local block,

$$\ell(b) = 1 + \lfloor (b-1)/p \rfloor \ , \quad b = 1, \ldots, q \ . \qquad (12.38)$$

With the definition of the remainder,

$$r = q \bmod p \ , \qquad (12.39)$$

and the ceiling function,

$$\ell_0 = \lceil q/p \rceil \ , \qquad (12.40)$$

the permutation operator for the cyclic distribution,

$$c = \pi(b) \ , \qquad (12.41)$$

obeys the formula,

$$\pi(b) = \begin{cases} (\alpha(b)-1)\ell_0 + \ell(b) & r = 0 \quad \alpha(b) = 1, \ldots, p \\ (\alpha(b)-1)\ell_0 + \ell(b) & r > 0 \quad \alpha(b) = 1, \ldots, r \\ (\alpha(b)-1)(\ell_0-1) + \ell(b) + r & r > 0 \quad \alpha(b) = r+1, \ldots, p \end{cases} \quad . \qquad (12.42)$$

The inverse permutation operator obeys the formula,

$$\pi^{-1}(c) = \alpha(c) + (\ell(c)-1)p \ . \qquad (12.43)$$

Table 12.1 shows an example of a cyclic distribution for the case $n = 13$, $p = 3$. The indices $\alpha(b)$ and $\ell(b)$ in columns two and three are the values defined by formulas (12.37) and (12.38). The values of the permutation operator

b	$\alpha(b)$	$\ell(b)$	$\pi(b)$	$\alpha(c)$	$\ell(c)$	$\pi^{-1}(c)$
1	1	1	1	1	1	1
2	2	1	6	1	2	4
3	3	1	10	1	3	7
4	1	2	2	1	4	10
5	2	2	7	1	5	13
6	3	2	11	2	1	2
7	1	3	3	2	2	5
8	2	3	8	2	3	8
9	3	3	12	2	4	11
10	1	4	4	3	1	3
11	2	4	9	3	2	6
12	3	4	13	3	3	9
13	1	5	5	3	4	12

TABLE 12.1: Cyclic distribution for the case $n = 13$ and $p = 3$.

shown in column four are defined by (12.42). The inverse permutation shown in column seven is the value from formula (12.43) using the indices for the permuted objects shown in columns five and six. The indices in the original set are dealt like cards to the images as shown in column seven. Image one holds five objects, numbered locally one through five as shown in column six, while images two and three hold only four objects, numbered locally one through four.

12.6 Load balancing

The use of subpartitions, sometimes called over-decomposition, is important for balancing the workload among images to prevent some images lying idle waiting for others. By creating more partitions than images and perhaps assigning partitions unequally among images, the programmer can balance the workload. Virtual-processor models assign work to processors dynamically with limited input required from the programmer [57] [58]. Although sometimes effective, such models remove a level of control from the programmer who knows better how to balance a workload than does a blind run-time system.

Molecular dynamics codes and particle-in-cell codes use over-decomposition to balance the workload. They decompose physical space into blocks and assign the blocks to images in a many-to-one fashion with the hope that as particles move in space from one block to another each image on average has the same amount of work to do. But particles move in unpredictable ways, and the workload tends to become unbalanced very quickly. In fact, if the particles accumulate in some partitions more than others, the programmer may need to reallocate the local blocks and reassign the particles. This reallocation generates a great deal of overhead if done too often, and special techniques may be required to obtain good performance for such codes [86] [87].

12.7   Exercises

1. Verify formula (12.33).

2. Verify that the operator R^q defined by Equation (12.36) and its transposed operator R_q define a partition of unity.

3. Verify formulas (12.42) and (12.43).

Chapter 13

Blocked Linear Algebra

Parallel application codes depend on maps between global indices and local indices either explicitly or implicitly. Chapter 12 described these maps as the composition of a subpartition operator with a permutation operator. The permutation operator changes the order of global block indices before the partition operator assigns them to local indices. The programmer transcribes formulas for these operations into computer code. These formulas are complicated, however, and transcribing them to computer code is subject to error.

This chapter describes a block matrix class that contains procedure pointers to functions that correctly implement these formulas. The programmer implements algorithms using these functions to map indices and can change the maps by associating the procedure pointers with different functions. This design demonstrates the powerful advantages of object-oriented design: the algorithms do not change, just the maps.

13.1 Blocked matrices

A matrix is a set of numbers labeled with two indices,

$$A = \{a_j^i, \quad i = 1, \ldots, m, \quad j = 1, \ldots, n\}, \tag{13.1}$$

the superscript labeling the rows and the subscript labeling the columns in a $[m, n]$ grid. Blocked matrices, used often for linear algebra algorithms, result from the application of two sets of subpartition operators independent of the image count. One set,

$$\sum_{J=1}^{M} Q_J Q^J = I_m, \tag{13.2}$$

cuts the row indices into M partitions with the sum of the operators equal to the identity operator I_m of dimension m, the global row dimension. Another set,

$$\sum_{K=1}^{N} R_K R^K = I_n \ , \tag{13.3}$$

cuts the column indices into N partitions with the sum of the operators equal to the identity operator I_n of dimension n, the global column dimension. Insertion of these identity operators left and right,

$$A = \sum_{J=1}^{M} \sum_{K=1}^{N} Q_J \left(Q^J A R_K \right) R^K \tag{13.4}$$

does not change the matrix but decomposes it into a sum of blocks,

$$A_K^J = Q^J A R_K \ , \quad J = 1, \ldots, M \ , \quad K = 1, \ldots, N \ , \tag{13.5}$$

where the upper case indices label blocks in a $[M, N]$ grid.

Formula (12.7), with appropriate substitution of parameters, yields the dimensions $(m(J), m(K))$ of block A_K^J, and formula (12.8) yields the base block indices $(k_0(J), k_0(K))$. Local block indices, therefore, correspond to global indices according to the formula,

$$
\begin{aligned}
(A_K^J)_k^j &= a_{k_0(K)+k}^{k_0(J)+j} \ , \\
J &= 1, \ldots, M \ , \quad K = 1, \ldots, N \ , \\
j &= 1, \ldots, m(J) \ , \quad k = 1, \ldots, m(K) \ .
\end{aligned}
\tag{13.6}
$$

The number of blocks is independent of how the blocks are assigned to images. The only requirement is that the partition numbers in each direction obey the inequalities,

$$1 \leq M \leq m \ , \quad 1 \leq N \leq n \ . \tag{13.7}$$

If $M = m$ and $N = n$, there is a single block of size (m, n). On the other hand, if $M = 1$ and $N = 1$, there are mn blocks each of size one. For $M > 1$ and $N = 1$, there are M blocks in the row direction but just one block in the column direction. In other words, it is a row-partitioned matrix. The case $M = 1$ for $N > 1$ yields a column-partitioned matrix. Some application codes specify the values for M and N explicitly. Other codes specify a preferred block size that determines the number of partitions based on the preferred size. The block size might be picked, for example, to fit blocks into cache or to minimize local or remote memory traffic.

Two block permutation operators, one for the M row blocks and an independent one for the N column blocks,

$$
\begin{aligned}
\Pi_M^T \Pi_M &= I_M \ , \\
\Pi_N^T \Pi_N &= I_N \ ,
\end{aligned}
\tag{13.8}
$$

permute the blocks into a different order,

$$A^{\pi_M(J)}_{\pi_N(K)} = \Pi_M A^J_K \Pi^T_N \ . \tag{13.9}$$

The maps from global indices to local indices yield the result,

$$(A^{\pi_M(J)}_{\pi_N(K)})^j_k = a^{k_0(J)+j}_{k_0(K)+k} \ , \quad j = 1,\ldots,m(J) \ , \quad k = 1,\ldots,m(K) \ , \tag{13.10}$$

for the local indices within each block. On the other hand, given the blocks in the permuted order,

$$(J', K') = (\pi_M(J), \pi_N(K)) \ , \tag{13.11}$$

the inverse map back to global indices obeys the formula,

$$(A^{J'}_{K'})^j_k = a^{k_0(\pi_M^{-1}(J'))+j}_{k_0(\pi_N^{-1}(K'))+k} \ , \quad j = 1,\ldots,m(\pi_M^{-1}(J')) \ , \quad k = 1,\ldots,m(\pi_N^{-1}(K')) \ . \tag{13.12}$$

Two sets of partition operators,

$$\sum_{\alpha=1}^r P_\alpha P^\alpha = I_M \ ,$$

$$\sum_{\beta=1}^s S_\beta S^\beta = I_N \ , \tag{13.13}$$

assign blocks to images with the constraint,

$$p = r \cdot s \ , \tag{13.14}$$

yielding a partition of the number of images p into a $[r, s]$ grid. Local block (J', K') assigned to image $[\alpha, \beta]$ obeys the formula,

$$\begin{aligned}
A^{\alpha,J'}_{\beta,K'} &= P^\alpha A^{\pi_M^{-1}(J')}_{\pi_N^{-1}(K')} S_\beta \\
J' &= 1,\ldots,\ell(\alpha) \ , \\
K' &= 1,\ldots,\ell(\beta) \\
\alpha &= 1,\ldots,r \ , \quad \beta = 1,\ldots,s \ ,
\end{aligned} \tag{13.15}$$

where the local indices on the right side are mapped back to the global indices through the inverse permutation. The map between local and global indices obeys the formula,

$$\begin{aligned}
\left[A^{\alpha,J'}_{\beta,K'}\right]^j_k &= a^{k_0(\pi_M^{-1}(b_0(\alpha)+J'))+j}_{k_0(\pi_N^{-1}(b_0(\beta)+K'))+k} \ , \\
J' &= 1,\ldots,\ell(\alpha) \ , \quad j = 1,\ldots,m(\pi_M^{-1}(J')) \ , \\
K' &= 1,\ldots,\ell(\beta) \ , \quad k = 1,\ldots,m(\pi_N^{-1}(K')) \ , \\
\alpha &= 1,\ldots,r \ , \quad \beta = 1,\ldots,s \ .
\end{aligned} \tag{13.16}$$

The result of all these manipulations is that the original matrix is the sum of its permuted blocks,

$$A = \sum_{\alpha=1}^{r} \sum_{J=1}^{M} \sum_{K=1}^{N} \sum_{\beta=1}^{s} Q_J \Pi_M^T P_\alpha \left[P^\alpha \Pi_M Q^J A R_K \Pi_N^T S_\beta \right] S^\beta \Pi_N R^K . \quad (13.17)$$

The programmer is free to choose the values for several parameters including the number of partitions of the matrix in each direction and independently the number of images in each direction of the image grid. In addition, the programmer can pick any permutation operator, including the identity permutation, with possibly different permutations for rows and columns.

Since the permutation matrices are orthogonal matrices, the transposed identities,

$$\begin{aligned}
\Pi_M \Pi_M^T &= I_M , \\
\Pi_N \Pi_N^T &= I_N ,
\end{aligned} \qquad (13.18)$$

hold just as well as the original identities (13.8). Swapping the roles of the permutation with its inverse permutation in identity (13.17), therefore, makes no difference in the formulas for the blocked data structures. The permutation is simply replaced by its inverse.

13.2   The block matrix class

Listing 13.2 shows a trucated version of the block matrix class. It is similar to the dense matrix class described in Listing 9.2 but at the same time different because the map between block indices and image indices is no longer a one-to-one map. The constructor function invoked as in Listing 13.1 accepts seven arguments, two more than the dense matrix class. The first two arguments specify the global row and column dimensions, the second two specify the number of blocks in the row and column directions, the third two specify the shape of the image grid, and the last argument specifies an optional halo width.

Listing 13.1: Invoking the constructor function.

```
A = blockMatrix(m,n,K,L,coDim1,coDim2,width)
```

The class also contains a set of procedure pointers to functions that represent the flurry of formulas from Section 13.1. When the constructor function creates an object of the class, it associates these pointers with appropriate functions that reside in separate library modules. The programmer uses these functions to map indices between global and local values and to map blocks to image indices. Forcing the programmer to use these functions reduces errors and provides flexibility to change the maps as long as the new maps conform to the interface blocks defined by the class. Separation of the index maps into libraries different from the libraries containing algorithms that use the maps means that algorithms do not change when the maps change.

Listing 13.2: The block matrix class.

```
module ClassBlockMatrix
  use ClassDenseMatrices
  implicit none
  integer,parameter, private :: wp    = kind(1.0d0)
  Type Matrix
    real(wp),allocatable :: a(:,:)
  end Type Matrix
  Type,extends(DenseMatrices) :: BlockMatrix
   integer :: globalRowBlocks    = 0
   integer :: globalRowRemainder = 0
   integer :: maxRowBlockDim     = 0
   integer :: globalColumnBlocks    = 0
   integer :: globalColumnRemainder = 0
   integer :: maxColumnBlockDim     = 0
   integer :: localBlocks    = 0
   integer :: maxLocalBlocks = 0
   integer :: haloWidth      = 0
   type(Matrix),allocatable :: block(:)
   integer,       allocatable :: iPivot(:,:)
!*************************Row functions*******************
  procedure(Interface1),pass(A),pointer &
   :: globalRowIndexToGlobalBlockIndex => null()   !(12.6r)
  procedure(Interface1),pass(A),pointer &
   :: globalRowBlockDim                 => null()   !(12.7r)
  procedure(Interface1),pass(A),pointer &
   :: globalRowBlockBaseIndex           => null()   !(12.8r)
  procedure(Interface1),pass(A),pointer &
   :: globalRowIndexToLocalRowIndex     => null()   !(12.10r)
  procedure(Interface2),pass(A),pointer &
   :: localRowIndexToGlobalRowIndex     => null()   !(12.11r)
  procedure(Interface1),pass(A),pointer &
   :: rowBlockIndexToFirstCoIndex    => null()   !(12.17r)
  procedure(Interface1),pass(A),pointer &
   :: localRowBlocks                 => null()   !(12.18r)
  procedure(Interface1),pass(A),pointer &
   :: localRowBlockBase              => null()   !(12.19r)
  procedure(Interface4),pass(A),pointer &
   :: globalRowBlockToLocalRowBlock => null()   !(12.21r)
  procedure(Interface3),pass(A),pointer &
   :: localRowBlockToGlobalRowBlock => null()   !(12.22r)
!*************************Column functions****************
  procedure(Interface1),pass(A),pointer &
```

```
    ::  globalColumnIndexToGlobalBlockIndex => null()  !(12.6c)
    procedure(Interface1),pass(A),pointer &
    ::  globalColumnBlockDim                => null()  !(12.7c)
    procedure(Interface1),pass(A),pointer &
    ::  globalColumnBlockBaseIndex          => null()  !(12.8c)
    procedure(Interface1),pass(A),pointer &
    ::  globalColumnIndexToLocalColumnIndex => null()  !(12.10c)
    procedure(Interface2),pass(A),pointer &
    ::  localColumnIndexToGlobalColumnIndex => null()  !(12.11c)
    procedure(Interface1),pass(A),pointer &
    ::  columnBlockIndexToSecondCoIndex     => null()  !(12.17c)
    procedure(Interface1),pass(A),pointer &
    ::  localColumnBlocks                   => null()  !(12.18c)
    procedure(Interface1),pass(A),pointer &
    ::  localColumnBlockBase                => null()  !(12.19c)
    procedure(Interface4),pass(A),pointer &
    ::  globalColumnBlockToLocalColumnBlock => null()  !(12.21c)
    procedure(Interface3),pass(A),pointer &
    ::  localColumnBlockToGlobalColumnBlock => null()  !(12.22c)
!*************Permutation functions**************
    procedure(Interface1),pass(A),pointer &
    ::  permuteRowIndex         => null()
    procedure(Interface1),pass(A),pointer &
    ::  inversePermuteRowIndex    => null()
    procedure(Interface1),pass(A),pointer &
    ::  permuteColumnIndex      => null()
    procedure(Interface1),pass(A),pointer &
    ::  inversePermuteColumnIndex => null()
!*************Global-local index maps*****************
    procedure(Interface5),pass(A),pointer &
    ::  globalBlockToLocalBlock => null()
    procedure(Interface5),pass(A),pointer &
    ::  localBlockToGlobalBlock => null()
!***************Linear algebra functions***************
    procedure(BlockMatVecInterface),pass(A),pointer &
    ::  blockMatVec     => null()
    procedure(MatMulInterface),pass(A),pointer &
    ::  matMul          => null()
    procedure(TransposeInterface),pass(A),pointer &
    ::  transpose       => null()
    procedure(LUInterface),pass(A),pointer &
    ::  LUDecomposition => null()
!****************Information functions******************
    procedure(Interface5),pass(A),pointer &
    ::  localBlockDims  => null()
    procedure(Interface5),pass(A),pointer &
    ::  localBlockBase  => null()
    procedure(Interface5),pass(A),pointer &
    ::  globalBlockDims => null()
```

```
  procedure(Interface5),pass(A),pointer &
  :: globalBlockBase => null()
  contains
!***************Destructor function********************
  final :: deleteBlockMatrix
  end Type BlockMatrix
!***************Constructor function*******************
  interface BlockMatrix
     procedure newBlockMatrix
  end interface BlockMatrix
!***************Interface blocks**********************
  interface
       :
  function Interface5(A,b) result(aj)
     import :: BlockMatrix
     Class(BlockMatrix),intent(in) :: A
     integer, intent(in) :: b(2)
     integer             :: aj(2)
  end function Interface5
       :
  end interface
```

Listing 13.3, for example, shows the function that maps two global block indices to an image index and a local block index. It conforms to the interface block shown in Listing 13.2. It first permutes the incoming block indices using two permutation functions, one for the row index and another for the column index. It then converts the permuted row and column indices into a grid of image indices and a grid of local block indices. It then linearizes the two image indices to a single image index based on the co-dimensions of the image grid and linearizes the local block indices into a single local index.

Listing 13.3: Mapping a global block to a local block.

```
function global_Block_To_Local_Block(A,JK) result(aj)
  Class(BlockMatrix),intent(in) :: A
  integer,intent(in)             :: JK(2)
  integer :: aj(2)
  integer :: J, K
  integer :: alpha, beta
  integer :: rows, columns
  integer :: rowIndex, columnIndex
  integer :: KL(2)
    J  = A%permuteRowIndex(JK(1))
    K  = A%permuteColumnIndex(JK(2))
    KL = A%globalRowBlockToLocalRowBlock(J)
    alpha    = KL(1)
    rowIndex = KL(2)
    KL = A%globalColumnBlockToLocalColumnBlock(K)
    beta        = KL(1)
    columnIndex = KL(2)
```

```
   rows   = A%localRowBlocks(alpha)
   aj(1)  = (beta-1)*A%coDim1 + alpha
   aj(2)  = (columnIndex-1)*rows + rowIndex
end function global_Block_To_Local_Block
```

Listing 13.4 shows the inverse function. It converts the linearized image index back to a grid of image indices and converts the linearized block index back to a grid of local block indices. Then it uses the inverse map to convert the local block indices back to the global block indices and finally applies the inverse permutation functions before returning the result.

Listing 13.4: Mapping a local block to a global block.

```
function local_Block_To_Global_Block(A,aj) result(JK)
   Class(BlockMatrix),intent(in)  :: A
   integer,intent(in)             :: aj(2)
   integer                        :: JK(2)
   integer :: alpha,beta
   integer :: i, j, k, p
     p = aj(1)
     k = aj(2)
     beta = (p-1)/A%coDim1 + 1
     alpha = p - (beta-1)*A%coDim1
     j     = (k-1)/A%localRowBlocks(alpha) + 1
     i     = k - (j-1)*A%localRowBlocks(alpha)
     JK(1) = A%localRowBlockToGlobalRowBlock([alpha,i])
     JK(2) = A%localColumnBlockToGlobalColumnBlock([beta,j])
     JK(1) = A%inversePermuteRowIndex(JK(1))
     JK(2) = A%inversePermuteColumnIndex(JK(2))
end function local_Block_To_Global_Block
```

Neither function knows anything about the permutation operators except that they conform to an interface block defined by the block matrix class. The constructor function associates the procedure pointers for the permutation functions when it creates an object. The programmer can change permutations by associating the pointers to whatever permutation function desired without changing the code for operations that depend on these permutations. In particular, the permutation can be swapped with its inverse by simply swapping the pointer assignments. Code for algorithms using the permutation functions does not change.

Listing 13.5: Block matrix-matrix multiplication.

```
function block_MatMul(A,B) result(C)
   Class(BlockMatrix),intent(in)  :: A
   Type(BlockMatrix),intent(in)   :: B
   Type(BlockMatrix),allocatable  :: C
   integer :: i,K,L,m,n
   integer :: dims(2), IJ(2), IK(2), KJ(2)
   integer :: ADims(2), BDims(2), CDims(2)
```

```
real(wp),allocatable :: tempA(:,:,:)[:]
real(wp),allocatable :: tempB(:,:,:)[:]
real(wp),allocatable :: bufferA(:,:)
real(wp),allocatable :: bufferB(:,:)
  L = A%maxLocalBlocks
  m = A%maxRowBlockDim
  n = A%maxColumnBlockDim
  allocate(tempA(m,n,L)[*])
  allocate(bufferA(m,n))
  L = B%maxLocalBlocks
  m = B%maxRowBlockDim
  n = B%maxColumnBlockDim
  allocate(tempB(m,n,L)[*])
  allocate(bufferB(m,n))
  do i=1,A%localBlocks
     dims = A%localBlockDims([A%me,i])
     tempA(1:dims(1),1:dims(2),i) &
        = A%block(i)%a(1:dims(1),1:dims(2))
  end do
  do i=1,B%localBlocks
     dims = B%localBlockDims([B%me,i])
     tempB(1:dims(1),1:dims(2),i) &
        = B%block(i)%a(1:dims(1),1:dims(2))
  end do
  m = A%globalRowDim
  n = B%globalColumnDim
  K = A%globalRowBlocks
  L = B%globalColumnBlocks
  C = BlockMatrix(m,n,K,L,A%coDim1,B%coDim2)
  sync all
  do i=1,C%localBlocks
     IJ = C%localBlockToGlobalBlock([C%me,i])
     do K=1,A%globalColumnBlocks
       IK = A%globalBlockToLocalBlock([IJ(1),K])
       KJ = B%globalBlockToLocalBlock([K,IJ(2)])
       ADims = A%localBlockDims([IK(1),IK(2)])
       BDims = B%localBlockDims([KJ(1),KJ(2)])
       CDims = C%localBlockDims([C%me,i])
       bufferA(1:ADims(1),1:ADims(2)) &
             = tempA(1:ADims(1),1:ADims(2),IK(2))[IK(1)]
       bufferB(1:BDims(1),1:BDims(2)) &
             = tempB(1:BDims(1),1:BDims(2),KJ(2))[KJ(1)]
       C%block(i)%a(1:ADims(1),1:BDims(2)) &
         = C%block(i)%a(1:ADims(1),1:BDims(2)) &
         + matmul(bufferA(1:ADims(1),1:ADims(2)) &
             , bufferB(1:BDims(1),1:BDims(2)))
     end do
  end do
  deallocate(tempA)
```

```
    deallocate(tempB)
end function block_MatMul
```

Listing 13.5 shows code for matrix-matrix multiplication that illustrates how to use these functions. The logic of the algorithm is the same as that shown in Listing 9.5 the difference being that there is no longer a one-to-one correspondence between block indices and co-dimension indices and the block indices have been permuted. Each image computes the result for blocks that it owns using blocks along a row in the first matrix of the product and blocks along a column in the second matrix of the product.

The matrix multiplication function uses four buffer arrays that hold copies of blocks from the two product matrices. Two of these arrays are co-array variables used as needed to move data from a remote image to the local image. They must be declared with the same size on each image with dimensions large enough to hold the largest block on any image. The allocation and deallocation statements for these arrays cause hidden synchronization. The other two buffer arrays are not co-array variables, and they are not essential to the logic of the algorithm. Remote blocks could be inserted as arguments directly into the call to the local matrix multiplication function. In that case, the compiler would generate its own temporary buffer arrays to hold the blocks. Performance may suffer, however, when the compiler is left to its own analysis.

13.3 Optimization of the LU-decomposition algorithm

The LU-decomposition algorithm attracts a great deal of attention because it is the basis for ranking computers in the Top 500 list [99]. The benchmark rules define a weak-scaling experiment with no specification for the weak-scaling constraint. Programmers grow the problem size to obtain the highest computational power for a single problem size for a single image count. The highest number of floating-point operations per second wins the prize.

Listing 13.6: Row block permutation functions.

```
function rowIndexPermutation(A,b) result(d)
   Class(BlockMatrix),intent(in) :: A
   integer, intent(in)           :: b
   integer                       :: d
   integer :: alpha, coDim, L, C, F, R
      coDim = A%coDim1
      R = mod(A%globalRowBlocks, coDim)
      C = (A%globalRowBlocks -1)/coDim + 1
      F = A%globalRowBlocks/coDim
      alpha = 1 + mod(b-1, coDim)
      L     = 1 + (b-1)/coDim
```

```
      if(alpha <= R) then
        d = (alpha-1)*C + L
      else if(alpha > R) then
        d = (alpha-1)*F + L  + R
      end if
end function rowIndexPermutation

function rowIndexInversePermutation(A,d) result(b)
   Class(BlockMatrix),intent(in)  :: A
   integer, intent(in)            :: d
   integer                        :: b
   integer :: alpha, coDim, L, C, F, R
     coDim = A%coDim1
     C = (A%globalRowBlocks-1)/coDim + 1
     F = A%globalRowBlocks/coDim
     R = mod(A%globalRowBlocks,coDim)
     if(d <= C*R) then
       alpha = (d-1)/C + 1
       L = d - (alpha-1)*C
     else if(d > C*R) then
       alpha = (d-1-R)/F + 1
       L = d - (alpha-1)*F - R
     end if
     b = alpha + coDim*(L-1)
end function rowIndexInversePermutation
```

The constructor function for the block matrix class, by default, associates procedure pointers with block-cyclic permutation functions. These permutation functions are the preferred choices for the LU-decomposition algorithm, but the programmer may want to make a different choice for another algorithm. Listing 13.6 shows the function that performs the cyclic permutation of a global row block index as well as the function that performs the corresponding inverse permutation. The permutation functions reside in a module separate from the module that defines the block matrix class, and the programmer may select different permutations by associating these pointers with different functions without changing the code for linear algebra algorithms. It is, of course, the programmer's responsibility to write valid permutation functions and to match them with the appropriate inverse functions. Performance of a particular algorithm may depend in subtle ways on the permutations used.

This flexibility in choosing permutation functions allows easy performance comparisons as the subpartition operators and the permutation operators change. Figure 13.1 shows results for a weak-scaling experiment where the global problem size grows according to the rule $n(p) = p^{3/4}n(1)$ where $n(1)$ is the initial size for one image. The lower curve, marked by triangles ($\triangle$), corresponds to one block assigned to each image with no subpartitions. The middle curve, marked with grads ($\triangledown$), adds subpartitions with many blocks assigned to each image but with no permutation employed. The top curve, marked with bullets (•), adds block-cyclic permutations and shows almost perfect scaling. The dashed line in the figure represents perfect linear scaling in the absence of communication overhead.

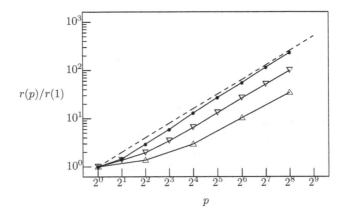

FIGURE 13.1: Computational power as a function of image count for LU decomposition: (•) weak scaling with block-cyclic permutation; ($\triangledown$) weak scaling with no permutation; ($\triangle$) weak scaling with no subpartitions. Results are shown relative to the value on one image.

These results confirm the efficacy of over-decomposition with permutations. Although the logic of the LU-decomposition code is the same as the logic displayed in the simpler code shown in Listing 9.7, the complications introduced by subpartition operators along with permutation operators makes the code harder to write and harder to maintain.

Following all the optimization techniques used to increase performance of the LU-decomposition algorithm has produced a cadre of programmers across the globe chasing all possible variations. There are many parameters to tune including the number of images employed, the global matrix size, the block sizes, and the shape of the image grid as outlined in some of the original papers [32] [33] [9] [36] [41]. Several papers describing performance models for LU-decomposition also exist [80] [81]. Some sticky issues arise generating large non-singular matrices from random number generators [31].

13.4   Exercises

1. Why is the `sync all` statement required in Listing 13.5?

2. Verify that the arguments supplied to the constructor function in Listing

13.5 yield a result matrix with all the conformal rules of matrix-matrix multiplication satisfied.

3. Modify the code shown in Listing 13.6 to perform a cyclic permutation of a global column block index.

4. Must the row and column permutations be the same? That is, if the rows are permuted using the cyclic permutation, can the columns be permuted using, for example, a random permutation?

5. Write a function for the identity permutation. Write a function that reverses the order of the block indices. What is the inverse function? Test that each is the inverse of the other.

6. Modify the code in Listing 13.5 by replacing the two buffers used in the local matrix multiplication operation with the corresponding co-array buffers.

7. Verify that the two functions shown in Listing 13.6 are inverses of each other.

8. Replace the block-cyclic permutation shown in Listing 13.6 with a random permutation. What changes need to be made to the code for the LU-decomposition algorithm?

Chapter 14

The Finite Element Method

The finite element method is a powerful technique for solving partial differential equations over irregular spatial domains. It is difficult to implement in parallel, and it requires a more sophisticated application of the co-array model than did previous examples. Fortunately, the book has already discussed all the basic techniques needed.

A finite element mesh consists of a set of nodes and a set of edges between nodes that together define a set of elements covering the spatial domain. A global index set labels the nodes, and a second global index set labels the elements. As in previous examples, permutation operators and partition operators shuffle and cut the global index sets, but different operators apply separately to nodes and elements. Because the set of edges defines a complicated relationship between the two index sets, these operators are more complicated than for previous examples, and the resulting maps between global and local indices are also more complicated.

No matter how cleverly the permutation and partition operators shuffle and cut the two global index sets, some elements assigned to one image unavoidably contain nodes assigned to another image, and conversely some nodes assigned to one image belong to elements assigned to another image. It is necessary, therefore, to define exchange operators that add halos around local meshes such that each image owns complete information about its mesh. From this local information, each image assembles mass and stiffness matrices in the sparse compressed-row-storage format. These matrices define a sparse system of equations, and an iterative solver yields the solution to the partial differential equation evaluated at the nodes of the mesh.

14.1   Basic ideas from finite element analysis

A typical example is the time-dependent heat equation defined on a two-dimensional spatial domain,

$$\frac{\partial u}{\partial t}(x,t) = \Delta u(x,t) + f(x) , \quad x \in \Omega \subset R^2 , \quad t \geq 0 . \tag{14.1}$$

The finite element method for solving this problem is based on the idea of a weak solution that minimizes a certain functional [2, ch. 5] [92]. In particular, given a function $f \in L_2(\Omega)$ and a space of well-behaved test functions $v \in V \subset L_2(\Omega)$, the weak solution obeys the equation,

$$\frac{\partial}{\partial t}(v, u) = (v, \Delta u) + (v, f) , \quad v \in V , \tag{14.2}$$

where (v, u) is the inner product over the space $L_2(\Omega)$. Integration by parts, yields the equation,

$$\frac{\partial}{\partial t}(v, u) = -(\nabla v, \nabla u) + (v, f) + (v, \nabla u)_{\partial\Omega} , \quad v \in V , \tag{14.3}$$

where $\partial\Omega$ is the boundary of the domain. With test functions set to zero on the boundary, the boundary term on the right side of (14.3) disappears.

An explicit time-stepping method, starting with an initial function $u^0(x) = u(x, 0)$ obtains the solution at the new time $u^{n+1}(x) = u(x, t_{n+1})$ from the solution at the previous time $u^n(x) = u(x, t_n)$ using, for example, the Euler method,

$$(v, u^{n+1}) = (v, u^n) - \Delta t \left[(\nabla v, \nabla u^n) - (v, f) \right] , \quad v \in V . \tag{14.4}$$

More elaborate time-stepping algorithms, of course, can be used.

The Ritz-Galerkin method expands the function in terms of a finite, time-independent basis set,

$$\phi_i(x) \in V , \quad i = 1, \ldots, N , \tag{14.5}$$

one for each of the N nodes of the mesh, such that

$$u(x,t) = \sum_{i=1}^{N} a_i(t)\phi_i(x) . \tag{14.6}$$

The time dependence shifts to the coefficients $a_i(t)$, and finding these coefficients solves the problem.

Since the functional Equation (14.4) must hold for all $v \in V$, it certainly holds for each basis function $v = \phi_i$. Replacing the function v in succession by the basis functions ϕ_i, yields a system of N equations in N unknowns,

$$\sum_{j=1}^{N} (\phi_i, \phi_j) a_j^{n+1} = \sum_{j=1}^{N} (\phi_i, \phi_j) a_j^n - \Delta t \left[\sum_{j=1}^{N} (\nabla \phi_i, \nabla \phi_j) a_j^n - F_i \right], i = 1, \ldots, N,$$

(14.7)

where $F_i = (\phi_i, f)$. The inner product between basis functions defines the mass matrix,

$$M_{i,j} = (\phi_i, \phi_j) ,$$

(14.8)

and the inner product between the derivatives of the basis functions defines the stiffness matrix,

$$K_{i,j} = (\nabla \phi_i, \nabla \phi_j) .$$

(14.9)

Starting from some initial vector of expansion coefficients $a^0 = a(t_0)$, the vector of coefficients at subsequent times $a^{n+1} = a(t_{n+1})$ is the solution of the system of equations,

$$M(a^{n+1} - a^n) = -\Delta t [K a^n - F] .$$

(14.10)

At each time step, the right side of (14.10) requires a matrix-vector multiplication by the stiffness matrix using the solution vector from the previous time step. The change in the solution to the next time step requires the solution of a sparse system of linear equations determined by the mass matrix. Saad gives an excellent description of modern methods for solving these sparse systems of equations [90].

For linear, time-independent problems like the heat equation, the matrices are constant in time and need to be formed just once. More complicated problems involving non-linear or time-dependent operators may require rebuilding the matrices at each time step [104].

14.2   Nodes, elements and basis functions

Figure 14.1 shows a pentagon covered by triangular elements specified by the coordinates of nodes at the corners of the triangles and by the edges between nodes [2, p. 167]. For modern application codes, the number of elements is typically very large. The number of nodes is also very large, but typically smaller than the number of elements because elements share nodes. The size of the resulting system of equations equals the number of nodes, and the nonzero entries in the sparse matrices depend on how the nodes are connected and on the choice of basis functions.

The key to the effectiveness of the finite element method is that the basis

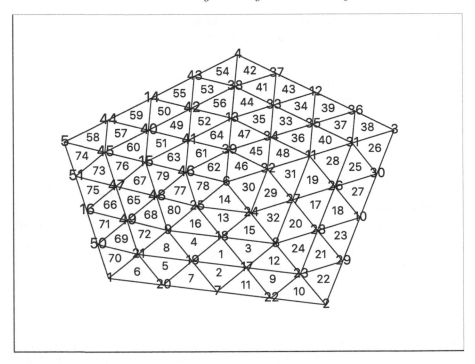

FIGURE 14.1: A finite element mesh over a pentagon with 80 triangular elements and 51 nodes at the corners of the triangles. Element indices mark the centers of the elements, and node indices mark the corners of the triangles.

function at node n_i is a sum over elemental functions,

$$\phi_i(x) = \sum_L \phi_i^L(x) , \quad L \in \{ \text{ element indices that share node } n_i \}, \quad (14.11)$$

where the index L extends over the elements that share the node. The elemental function $\phi_i^L(x)$ is a low-order polynomial, often just a linear function, constructed equal to one at node n_i and equal to zero along the edge determined by the other nodes of element L. The basis function obeys the property,

$$\phi_i(x_j) = \delta_j^i , \quad (14.12)$$

where the Kronecker delta equals one when $x_j = x_i$ at the location of node n_i and equals zero at $x_j \neq x_i$ at the location of any other node n_j different from node n_i. Substitution of (14.12) into Equation (14.6), then, yields the value of the solution at node n_i,

$$a_i(t_n) = u(x_i, t_n) , \quad (14.13)$$

at each time step.

With linear elemental functions, basis function ϕ_i forms a tent centered over node n_i as shown in Figure 14.2. The support of the basis function is the subset of elements that share node n_i. Outside this subset, the function is zero. Two basis functions, therefore, have nonzero overlap only between nodes that share a common element.

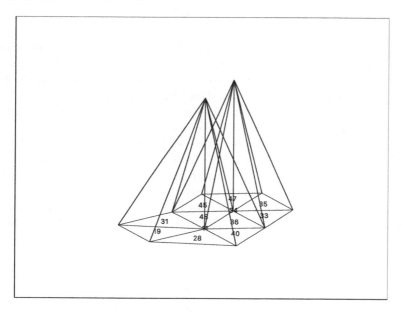

FIGURE 14.2: Overlapping tent functions centered over nodes n_{11} and n_{34} for the mesh shown in Figure 14.1. The functions have nonzero overlap only between the shared elements e_{36} and e_{48}.

The value of the mass matrix element (14.8) corresponding to the interaction between nodes n_i and n_j is a double sum,

$$M_{ij} = \left(\sum_L \phi_i^L, \sum_{L'} \phi_j^{L'}\right) = \sum_L \sum_{L'} (\phi_i^L, \phi_j^{L'}) \tag{14.14}$$

over elements that share both nodes. By construction, the elemental functions obey the orthogonality relationship,

$$(\phi_i^L, \phi_j^{L'}) = \delta_L^{L'} (\phi_i^L, \phi_j^L) \tag{14.15}$$

where $\delta_L^{L'}$ is the Kronecker delta, equal to one if the elements are the same and equal to zero if they are not the same. The double sum, therefore, collapses to a single sum over elemental matrices,

$$M_{ij} = \sum_L M_{ij}^L , \tag{14.16}$$

where

$$M_{ij}^L = (\phi_i^L, \phi_j^L) \,, \quad L \in \{ \text{ element indices that share nodes } n_i \text{ and } n_j \ \} \,.$$
$$(14.17)$$

Similarly, the stiffness matrix is the sum of elemental stiffness matrices,

$$K_{ij} = \sum_L K_{ij}^L \,. \tag{14.18}$$

The mass matrix and the stiffness matrix, therefore, are sparse symmetric matrices with dimension equal to the number of nodes and with non-zero entries only between nodes that share a common element. The value of the off-diagonal element for the example shown in Figure 14.2 is the sum over two shared elements. The values of the diagonal elements, on the other hand, are sums over all six elements shared by the node.

Matrix assembly takes place in two steps. The first step computes the local contribution from each element determined by the overlap of the elemental functions. For triangular elements, there are three such functions, and they overlap to form local three-by-three contributions to the global matrix. These matrices are sparse with non-zero entries labeled by the global node indices. With zero entries compressed out, the elemental mass matrix $\hat{M}^L$ for element e_L has the value,

$$\hat{M}^L = \frac{\mid A^L \mid}{12} \cdot \begin{bmatrix} 2 & 1 & 1 \\ 1 & 2 & 1 \\ 1 & 1 & 2 \end{bmatrix} \,. \tag{14.19}$$

These matrices depend on the area of the triangle that defines a particular element,

$$A^L = (1/2) \cdot \left[(x_j^L - x_i^L)(y_k^L - y_i^L) - (x_k^L - x_i^L)(y_j^L - y_i^L) \right] \,, \tag{14.20}$$

computed in terms of the coordinates of the three nodes that define the triangle [27]. A well-defined mesh requires these areas to be about the same for all elements. Similarly, the local elemental stiffness matrices,

$$\hat{K}^L = \frac{1}{4 \mid A^L \mid} \cdot (D^L)^T D^L \,, \tag{14.21}$$

again with the zero entries compressed out, are determined by the coordinates of the triangle through the matrix [27],

$$D^L = \begin{bmatrix} x_k^L - x_j^L & x_i^L - x_k^L & x_j^L - x_i^L \\ y_k^L - y_j^L & y_i^L - y_k^L & y_j^L - y_i^L \end{bmatrix} \,. \tag{14.22}$$

The physical coordinates of the nodes, therefore, determine the values of the entries in the sparse matrices that define the finite element method.

The second step in the assembly process sums together these local contributions to obtain the full value for each entry in the global matrix with the three local node indices expanded back to the global node indices. Efficient implementation of this complicated assembly process, for both sequential and parallel codes, is a topic of current research [2, ch. 5] [27] [28] [55] [90, ch. 2].

14.3 Mesh partition operators

Implementation of a parallel finite element code is yet another exercise in defining maps between index sets. It is more difficult than previous examples because there are three index sets. One set labels the elements and another set labels the nodes. The two sets are related to each other, but they must be partitioned independently and they must be assigned independently to a third set that labels the images. Although the process is complicated, describing it in terms of partition and permutation operators puts the finite element problem into the same general framework as previous examples.

Without permutations, partition operators that simply cut the index sets in the order listed take no account of how the elements and nodes are related to each other. The result, most likely, is a disconnected set of elements and a disconnected set of nodes assigned to the image index set resulting in unnecessary movement of data between images. Permutation operators renumber the elements and nodes in some coherent way before they are partitioned to minimize communication and to balance the workload.

Finding an optimal partitioning algorithm is a very difficult problem, but fortunately a number of publicly available software packages address the problem such as the PT-Scotch Library [25], the Chaco Library [48], and the Metis Library [59]. These packages perform heuristic algorithms that take a description of an existing mesh as input and produce two lists as output, one that defines the elements assigned to each image and another that defines the nodes assigned to each image. Because the algorithms are different for different packages, they may produce different partitions for the same mesh.

Altough the details of the different packages may differ, they all start with a mesh defined by a set of elements,

$$\mathcal{E} = \{e_1, \ldots, e_M\} \, , \tag{14.23}$$

labeled with a set of global indices, $1 \leq i \leq M$. The algorithms permute the elements to a different order,

$$\Pi \mathcal{E} = \{e_{\pi(1)}, \ldots, e_{\pi(M)}\} \, , \tag{14.24}$$

and then assign the newly ordered elements to images,

$$P^\alpha \Pi \mathcal{E} = \{e_1^\alpha, \ldots, e_{M_\alpha}^\alpha\} \, , \tag{14.25}$$

where the partition operators P^α are similar to previous operators with the difference that the assignment to images and the local partition size M_α are decided by the package used to partition the mesh.

Similarly, the nodes,

$$\mathcal{N} = \{n_1, \ldots, n_N\} \, , \tag{14.26}$$

are labeled by an independent set of indices $1 \leq i \leq N$. The mesh-partition algorithms produce a different permutation operator,

$$\Lambda \mathcal{N} = \{n_{\lambda(1)}, \ldots, n_{\lambda(N)}\} , \tag{14.27}$$

and a different partition operator,

$$R^\alpha \Lambda \mathcal{N} = \{n_1^\alpha, \ldots, n_{N_\alpha}^\alpha\} , \tag{14.28}$$

that assigns the nodes independently to images. The two steps define the partition operations as composite operations, one for the elements,

$$Q^\alpha = P^\alpha \Pi , \tag{14.29}$$

and one for the nodes,

$$S^\alpha = R^\alpha \Lambda . \tag{14.30}$$

These composite operations are related to the Composite Pattern in object-oriented design [39].

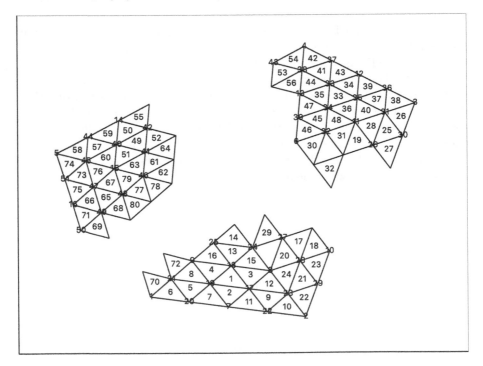

FIGURE 14.3: The mesh from Figure 14.1 cut into three partitions by the Metis package [59].

Figure 14.3 shows the mesh from Figure 14.1 partitioned into three pieces by the Metis package [59]. Notice that elements along the cuts assigned to one image may not own all the nodes that define the element. And some elements might better be assigned to different images. To assemble the mass and stiffness matrices, images need information from all the elements shared by a node and from each node that defines an element. Part of a finite element implementation, therefore, involves an exchange operation. The exchange operations are more complicated than the ones described in Chapter 11, because the exchange takes place between irregular partners rather than between nearest neighbors.

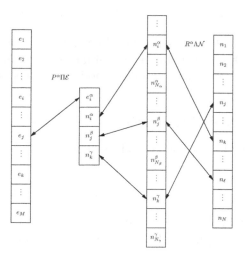

FIGURE 14.4: The relationship between global elements mapped to local elements and global nodes mapped to local nodes. The double arrows imply that the elements and nodes remain the same while the indices used to label them change. The double arrows also imply that there is an invertible map between global and local indices defined by the operators Q^α from definition (14.29) and S^α from (14.30).

Figure 14.4 illustrates some of the complications that arise while implementing a parallel version of a finite element code. The elements and nodes are intimately connected with each other, but they are distributed differently across images. The elements on the left in the figure are labeled by global element indices, while the nodes on the right are labeled by global node indices. Before permutation and partition, each element is defined by three nodes,

$$e_i = e_i(n_a, n_b, n_c) \ , \quad i = 1, \ldots, M \ . \tag{14.31}$$

labeled with global node indices, and each node is described by a list of elements that share the node,

$$n_i = n_i(e_s, e_t, \ldots, e_u, e_v) \ , \quad i = 1, \ldots, N \ , \tag{14.32}$$

labeled with global element indices. For a sequential code, this information is enough to implement the method.

For a parallel implementation, the labels are mixed up and the code must keep track of how they are mixed. One permutation operator assigns the elements to images,

$$(Q^\alpha \mathcal{E})_j = e_j^\alpha(n_a, n_b, n_c) \,, \quad j = 1, \ldots, M^\alpha \,, \tag{14.33}$$

with the same list of nodes still labeled by the global node indices. A second permutation operator independently assigns nodes to images,

$$(S^\alpha \mathcal{N})_j = n_j^\alpha(e_s, e_t, \ldots, e_u, e_v) \,, \quad j = 1, \ldots, N^\alpha \,, \tag{14.34}$$

with the same list of elements still labeled by the global element indices. The parallel code, therefore, must convert global indices to local indices such that each local element knows how to find its nodes in terms of local indices on remote images,

$$e_i^\alpha = e_i^\alpha(n_a^\beta, n_b^\gamma, n_c^\delta) \,, \quad i = 1, \ldots, M^\alpha \,, \tag{14.35}$$

and each local node knows how to find its elements in terms of local indices on remote images,

$$n_j^\alpha = n_j^\alpha(e_s^\beta, e_t^\gamma, \ldots, e_u^\delta, e_v^\epsilon) \,, \quad j = 1, \ldots, N^\alpha \,. \tag{14.36}$$

These requirements yield complicated code for parallel implementations of the finite element method, and the programmer must pay careful attention to detail.

14.4   The mesh class

A well-designed mesh class mitigates the difficulties encountered in the design and implementation of a parallel finite element code. A detailed description of the class is long and complicated and adds little to understanding the class. It is enough to show a truncated version of the class as in Listing 14.1 with the remark that it resembles the familiar design of previous classes. It contains a list of node objects and a list of element objects that define the mesh, and a set of procedure pointers. The constructor functions set these pointers to functions that exchange nodes and elements, that build the mass and stiffness matrices, and that map indices local-to-global and global-to-local.

Listing 14.1: The mesh class.

```
module ClassMesh
 use ClassNode
 use ClassElement
 use ClassCSRMatrix
```

```
use Collectives
implicit none
integer, private,parameter :: wp   = kind(1.0d0)
Type Mesh
  integer :: p
  integer :: me
  integer :: numberOfNodes
  integer :: numberOfElements
  integer :: localElements
  integer :: ghostElements
  integer :: localNodes
  integer :: ghostNodes
  integer, allocatable :: elementsToNodes(:,:)
  real(wp),allocatable :: coordinates(:,:)
  logical, allocatable :: bndryNode(:)
  integer,allocatable  :: Lambda(:)
  integer,allocatable  :: LambdaInverse(:)
  Type(Element),allocatable :: elements(:)
  Type(Node),   allocatable :: nodes(:)
  procedure(NonZerosInterface),pointer,pass(M)   &
   :: nonZeros => null()
  procedure(distRealInterface),pointer,pass(M)   &
   :: distributeRealDataFile => null()
  procedure(distIntegerInterface),pointer,pass(M) &
   :: distributeIntegerDataFile => null()
  procedure(distLogicalInterface),pointer,pass(M) &
   :: distributeLogicalDataFile => null()
  procedure(GlobalIndexToLocalInterface),pointer,pass(M) &
   :: globalDataIndexToLocalDataIndex => null()
  procedure(MassInterface),pointer,pass(M)       &
   :: MassMatrix => null()
  procedure(MassInterface),pointer,pass(M)       &
   :: AdjacencyMatrix => null()
  procedure(StiffInterface),pointer,pass(M)      &
   :: StiffnessMatrix => null()
  procedure(N2EInterface),pointer,pass(M)        &
   :: nodeToElement => null()
  procedure(N2NInterface),pointer,pass(M)        &
   :: nodeToNode    => null()
  procedure(GEToLEInterface),pointer,pass(M)     &
   :: globalElementToLocalElement => null()
  procedure(GNtoLNInterface),pointer,pass(M)     &
   :: globalNodeToLocalNode => null()
  procedure(BNodeInterface),pointer,pass(M)      &
   :: isBoundaryNode => null()
  procedure(exNodeInterface),pointer,pass(M)     &
   :: exchangeGhostNodes => null()
  procedure(exElemInterface),pointer,pass(M)     &
   :: exchangeGhostElements => null()
```

```
contains
  final :: deleteMesh
end Type Mesh
  interface Mesh
    procedure :: newMesh1
    procedure :: newMesh2
  end interface Mesh

  interface
    :
  end interface
contains
 recursive function newMesh1(Min,lvls) result(M)
  Type(Mesh),intent(in)        :: Min
  integer,intent(in),optional :: lvls
  Type(Mesh),allocatable      :: M
    :
 end function newMesh1
 function newMesh2(Elem2Node,Coord,Node2Elem,    &
                   Node2Node,ElemPart,NodePart,  &
                   Boundary) result(M)
  Class(Mesh),allocatable :: M
  character(len=*),intent(in) :: Elem2Node, Coord
  character(len=*),intent(in) :: Node2Elem, Node2Node
  character(len=*),intent(in) :: ElemPart,  NodePart
  character(len=*),intent(in) :: Boundary
    :
 end function newMesh1
```

One form of the constructor function accepts an existing mesh object as input and recursively refines it. Listing 14.2, for example, shows code that constructed the mesh displayed in Figure 14.1. As the code shows, the input mesh is specified by the node coordinates, the element edges, and a list of the boundary nodes. In this example, the initial mesh consists of five nodes defined by the coordinates at the corners of the pentagon plus the coordinates of the centroid of the pentagon and by a list of element edges specifying how the nodes are connected to form triangles. At each level of refinement, the constructor cuts the existing triangles into four new triangles. It adds nodes halfway between each corner, records the coordinates of the new nodes, updates the edges to form new elements, and records the new boundary nodes. After one level of refinement, the original mesh with five elements becomes a mesh with $4 \times 5 = 20$ elements, and after two levels of refinement it becomes a mesh with $4 \times 4 \times 5 = 80$ elements as shown in Figure 14.1.

Listing 14.2: Refining an existing mesh.

```
Type(Mesh) :: M
    :
! Node coordinates
```

```
 N(1,1:2)  =  [1.0 ,    -3.5]
 N(2,1:2)  =  [3.5 ,    -4.5]
 N(3,1:2)  =  [4.3 ,     1.9]
 N(4,1:2)  =  [2.5 ,     4.7]
 N(5,1:2)  =  [0.5 ,     1.5]
 N(6,1:2)  =  [sum(N(1:5,1))/5.0_wp,sum(N(1:5,2))/5.0_wp]
! Element edges
 E(1,1:3)  =  [1,2,6]
 E(2,1:3)  =  [2,3,6]
 E(3,1:3)  =  [3,4,6]
 E(4,1:3)  =  [4,5,6]
 E(5,1:3)  =  [1,5,6]
! Boundary nodes
 bndry(1:5)  =  .true.
 bndry(6)    =  .false.
! Inital mesh
 M%coordinates     = N
 M%elementsToNodes = E
 M%bndryNode       = bndry
! Refine the mesh
 M = Mesh(M,levels)
```

A mesh object has pointers to functions that return different ways to represent the mesh. One function returns a list of node coordinates. Another function returns a list of edges, that is, how each node is connected to its neighbors. Yet another function returns a list of nodes that define the three corners of each element, and another function returns a list of elements that share each node. All this information, plus a list of boundary nodes, can be written to files that become input for a second form of the constructor function.

Listing 14.3: A second constructor function for a mesh object.

```
 M = Mesh(file1,file2,file3,file4,file5,file6,file7)
```

Listing 14.3 shows the invocation of the alternative form of the constructor function. It creates a partitioned mesh distributed across images from information contained in seven files. The first four files contain an element-to-node map, a coordinate map, a node-to-element map, and a node-to-node map. The fifth file contains an element-to-image map, and the sixth file contains a node-to-image map generated by a mesh-partition algorithm, for example, by the Metis library [59]. The seventh file contains a boundary map.

The constructor function reads the files and distibutes the information across images. Each image independently searches for the elements and nodes assigned to it, and constructs a list of local nodes and local elements by invoking node and element constructor functions. Each image invokes exchange procedures to fill in the ghost nodes and ghost elements needed to expand the mesh to a self-contained local mesh. These exchange operators are generalizations of the halo-exchange operations described previously in Chapter 11.

Figure 14.5 shows the result of adding ghosts to the partitioned mesh shown previously in Figure 14.3.

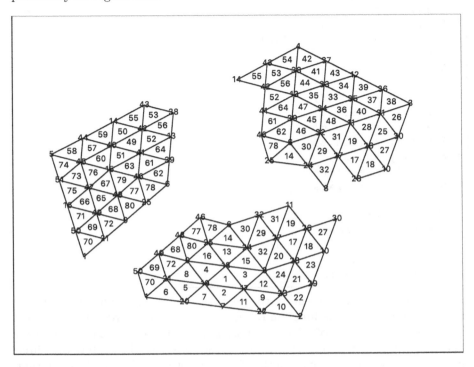

FIGURE 14.5: Ghost nodes and ghost elements added along the cuts in the original mesh. Each image owns a complete local mesh.

14.5   Integrating the heat equation

An example that shows how to solve partial differential equations is more useful than looking further into the details of the mesh class. Integration of the heat equation using the Euler method,

$$M(a^{n+1} - a^n) = -\Delta t \, [Ka^n - F] \, , \tag{14.37}$$

provides a simple example. Two functions associated with the mesh object create the mass and stiffness matrices as compressed-sparse-row (CSR) matrices described in Chapter 7. All the sparse linear algebra developed for these matrix objects is available to integrate the matrix equation. At each time step, a sparse matrix-vector multiplication on the right side is followed by the solution of a sparse system on the left side.

A number of technical details arise in the parallel implementation. The partition operator (14.28) cuts the rows of the matrices,

$$M^\alpha = R^\alpha \Lambda M , \quad K^\alpha = R^\alpha \Lambda K , \quad F^\alpha = R^\alpha \Lambda F , \tag{14.38}$$

to yield a partitioned system of equations,

$$M^\alpha(a^{n+1} - a^n) = -\Delta t \left[K^\alpha a^n - F^\alpha \right] . \tag{14.39}$$

Since the columns have not been permuted, the matrices are no longer symmetric, and the conjugate gradient algorithm no longer applies. Because the algorithm used to solve sparse equations is represented by a procedure pointer, however, a switch to another solver, such as the parallel bi-conjugate-gradient-stabilized (biCGStab) algorithm, fixes this problem.

This problem can be avoided by permuting the columns of the matrix to maintain symmetry,

$$M^\alpha = R^\alpha(\Lambda M \Lambda^T) , \quad K^\alpha = R^\alpha(\Lambda K \Lambda^T) , \quad F^\alpha = R^\alpha \Lambda F , \quad b^n = \Lambda a^n , \tag{14.40}$$

yielding the symmetric sparse system of equations,

$$M^\alpha(b^{n+1} - b^n) = -\Delta t \left[K^\alpha b^n - F^\alpha \right] . \tag{14.41}$$

The column permutation operation requires only a permutation of the column index array at the end of the functions that create the matrices. The permutation and partition of these sparse matrices are of the same kind as the permutation and partition of dense matrices displayed in definition (13.15).

Listing 14.4: Integrating the heat equation.

```
subroutine solveHeat(M)
  use ClassMesh
  use ClassCSRMatrix
  use ClassSolver
  use Collectives, only : sumToAll, minToAll
  implicit none
  Type(Mesh),intent(in)  :: M
  integer,parameter      :: wp      = kind(1.0d0)
  integer,parameter      :: itMax = 20
  integer                :: n
  real(wp),parameter     :: zero  = 0.0_wp
  real(wp),parameter     :: one   = 1.0_wp
  real(wp),parameter     :: eps   = 1.0d-07
  Type(CSRMatrix)        :: Mass
  Type(CSRMatrix)        :: Stiff
  real(wp),allocatable   :: a(:),da(:)
  real(wp),allocatable   :: f(:),g(:),force(:)
  real(wp)               :: area,dt,t,tMax
    area = minval(M%elements(1:M%localElements)%area)
```

```
      area = minToAll(area)
      Mass  = M%MassMatrix()
      Stiff = M%StiffnessMatrix()
      allocate(a(M%numberOfNodes))
      allocate(da(M%numberOfNodes))
      allocate(force(M%numberOfNodes))
      n = M%localNodes
      allocate(f(n))
      allocate(g(n))
      a     = zero
      da    = zero
      force = zero
      dt    = area**3
      tMax  = 50*dt

   !  Set the force values
              :
      f(1:n) = force(Mass%K_0+1:Mass%K_0+n)

   !  Euler integration
      t = zero
      do while(t < tMax)
         g = -dt*(Stiff%matVec(a)-f)
         da(Mass%K_0+1:Mass%K_0+n) = Mass%solve(g,itMax,eps)
         a = a + sumToAll(da)
         t = t + dt
      end do
end subroutine solveHeat
```

Listing 14.4 shows code for the integration of the heat equation. It incorporates almost all the basic techniques developed throughout the book and illustrates how to develop a parallel application using object-oriented design combined with the co-array programming model. Figure 14.6 shows the solution of the heat equation starting from an initial value equal to zero with zero Dirichlet conditions on the boundary. The external force equals zero on the boundary but equals one across the internal nodes of the mesh. It pushes the temperature to positive values as time increases shown by bullets in the figure.

14.6   Exercises

1. Modify the code shown in Listing 14.2 to produce a mesh covering an octagonal domain. After three levels of refinement, how many elements does the mesh contain?

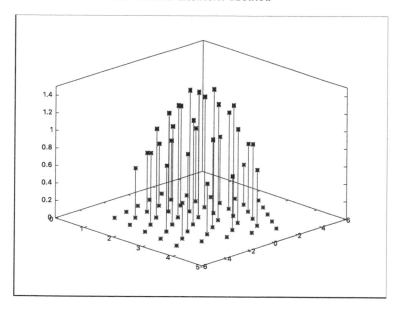

FIGURE 14.6: A solution to the heat equation with zero initial value, a positive external heat source, and zero boundary conditions. The temperature rises from zero above each interior node of the mesh.

2. Compare the elements assigned to images shown Figure 14.3 with the elements asigned to images after the exchange operation shown in Figure 14.5. Trace the ghost cells added to each image in the second figure and determine where they came from.

3. What changes must be made to the parallel conjugate gradient algorithm to convert it to a parallel biCGStab algorithm?

4. Integrate the heat equation using the Crank-Nicolson method,

$$[M + (\Delta t/2)K]a^{n+1} = [M - (\Delta t/2)K]a^n + \Delta t F .$$

Chapter 15

Graph Algorithms

Graphs provide convenient descriptions of important problems in science, engineering, and data analysis. Nodes of a graph represent items of interest and edges between nodes represent interactions between items. Graph theory and graph algorithms can provide important information about these interactions.

Because graphs and sparse matrices are related, previously defined distributed data structures designed for sparse matrix algorithms apply also to graph algorithms. The breadth-first search algorithm considered in this chapter is an example of how techniques designed for numerical algorithms relate to seemingly different non-numerical graph algorithms.

15.1 Graphs

A graph $G = (\mathcal{N}, \mathcal{E})$ consists of a set of nodes labeled by a set of integer indices,

$$\mathcal{N} = \{n_i\} , \quad i = 1, \dots, n ,$$

(15.1)

and a set of edges,

$$\mathcal{E} = \{e_{ij}\} ,$$

(15.2)

with edge e_{ij} connecting node $n_i \in \mathcal{N}$ with node $n_j \in \mathcal{N}$. Specification of a graph, therefore, is similar to the specification of a mesh without the restriction that the edges form triangles and without node coordinates that tie nodes to positions in space.

i	e_{ij}										
1	2										
2	1	3	6	7	8	9	11	12	14	15	16
3	2	16									
4	—										
5	14										
6	2	15									
7	2	12									
8	2	9	15								
9	2	8	15								
10	14										
11	2	12	14	15							
12	2	7	15	16							
13	—										
14	2	5	10	11	15						
15	2	6	8	9	11	12	14				
16	2	3	12								

$$(15.3)$$

Tableau (15.3) shows a set of edges for a graph with sixteen nodes. A node may have an edge connected to itself, called a self-loop, sometimes important for a particular problem sometimes not. Some nodes may be isolated with no connections to other nodes. If the set of edges contains edge e_{ij} but not necessarily edge e_{ji} in the opposite direction, the graph is called a directed graph. If the set contains both edges, the graph is undirected. The following discussion includes undirected graphs only.

Figure 15.1 shows a more informative description of the graph represented by the set of edges listed in tableau (15.3). Each node is marked by its index label and edges are represented as lines between nodes. These lines correspond to the entries on each line of tableau (15.3). Node two has eleven entries, and eleven lines radiate from node two in Figure 15.1. On the other hand, nodes four and thirteen are isolated with no connections to other nodes.

A one-to-one correspondence exists between a graph G with n nodes and a sparse matrix A of order n. The matrix is known as the adjacency matrix with elements,

$$a_{ij} = \begin{cases} 1 & e_{ij} \in \mathcal{E} \\ 0 & e_{ij} \notin \mathcal{E} \end{cases}.$$ $$(15.4)$$

The adjacency matrix for an undirected graph is symmetric. Conversely, for any matrix of order n, there is a corresponding graph with n nodes and a set of edges defined by entries in the matrix,

$$e_{ij} \in \mathcal{E}, \quad a_{ij} \neq 0$$
$$e_{ij} \notin \mathcal{E}, \quad a_{ij} = 0$$ $$(15.5)$$

If the matrix is symmetric, it corresponds to an undirected graph. If its nonzero elements have values other than one, it represents a weighted graph.

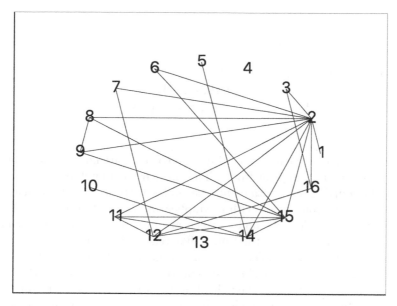

FIGURE 15.1: The sixteen-node graph corresponding to the set of edges shown in tableau (15.3). The nodes are located along a circle for convenience with no implications implied by their positions in space. Only the edges define connections between nodes.

15.2   The breadth-first search

The breadth-first-search algorithm consists of finding a sequence of disjoint sets of nodes such that each node in a set shares an edge with at least one node in the previous set. The sets are called fronts because they stretch across the graph from an arbitrary start node. The first front f^1 contains just the start node; the second front f^2 contains the neighbors of the start node; subsequent fronts $f^\ell, \ell = 3, 4, \ldots$ contain all the nodes with neighbors in front $f^{\ell-1}$ but not in lower fronts $f^{\ell-2}, \ldots, f^1$. Eventually, there are no new neighbors because the number of nodes is finite and the fronts are disjoint. The last front, therefore, is the null set.

For the graph shown in Figure 15.1 starting, for example, from node five there is only one edge connecting the start node to another node, to node fourteen. The start node is in front f^1, and node fourteen is in front f^2 as shown in equation 15.6.

$$\begin{aligned}
f^1 &= \{5\} \\
f^2 &= \{14\} \\
f^3 &= \{2, 10, 11, 15\} \\
f^4 &= \{1, 3, 6, 7, 8, 9, 12, 16\} \\
f^5 &= \{\emptyset\}
\end{aligned} \tag{15.6}$$

Node fourteen has five edges, including one to the start node, which must be excluded to maintain disjoint fronts. The third front, therefore, contains four nodes. The fourth front contains all the nodes with an edge to at least one node in the third front, again with no duplicates. The fifth front is empty and the search ends. Nodes four and thirteen are disconnected from other nodes and do not belong to any of the fronts that rediate from node five.

A node in one front is called a parent of a node in the next front if there is an edge connecting them. Picking a parent for each node in a front determines a path from one front to the next leading from the start node to any other node or conversely from any node back to the start node. Since a node may have more than one parent, the paths are not unique. A record of the parents picked at each front, however, is enough to describe the chosen paths. The start node records itself as parent. All nodes with the start node as parent record the start node. All children of these nodes record their parent and so forth. An isolated node records a zero indicating it has no parent, and no other node records it as a parent.

This record is called the parent array for a particular path from a start node to any other node. Array (15.7), for example, shows a parent array from start node five for the graph shown in Figure 15.1.

i	p_i	ℓ_i	i	p_i	ℓ_i
1	2	4	9	2	4
2	14	3	10	14	3
3	2	4	11	14	3
4	0	0	12	2	4
5	5	1	13	0	0
6	2	4	14	5	2
7	2	4	15	14	3
8	2	4	16	2	4

$$\tag{15.7}$$

It is not unique. Node six, for example, could pick node fifteen as its parent rather than node two.

The length of a path between two nodes is the difference between their front indices. The length of the path from node six to node five, for example, is three following the path,

$$n_6 \leftrightarrow n_2 \leftrightarrow n_{14} \leftrightarrow n_5 \ .$$

The path in the reverse direction, of course, is the same.

Finding nodes from one front to the next is equivalent to multiplying a sparse vector by a sparse matrix followed by elimination of duplicates. Let the set,

$$f^\ell = \{n_1^\ell, \ldots, n_{m^\ell}^\ell\} . \tag{15.8}$$

contain the m^ℓ nodes in front ℓ. For each node in the set, form a vector with zeros everywhere except for a value one at the row equal to the node index,

$$e_{n_j^\ell}^k = \delta_{n_j^\ell}^k , \quad n_j^\ell \in f^\ell , \quad j = 1, \ldots, m^\ell , \quad k = 1, \ldots, n . \tag{15.9}$$

Multiplication by the adjacency matrix (15.4) is the sum,

$$g_{n_j^\ell} = \sum_{k=1}^n A_k e_{n_j^\ell}^k \tag{15.10}$$

where A_k is the k-th column of the matrix. Only one column survives,

$$g_{n_j^\ell} = A_{n_j^\ell} , \tag{15.11}$$

with nonzero entries only for edges connected to the parent node in the front. Summing the results from each node in the front,

$$g^\ell = \sum_{j=1}^{m^\ell} g_{n_j^\ell} \tag{15.12}$$

yields a vector with nonzeros corresponding to candidate nodes for the next front. If an element of the vector g^ℓ equals one, the candidate node has just one parent. If an element of the vector is greater than one, the candidate has more than one possible parent. Squeezing out the zeros yields a set of candidate nodes,

$$h^\ell = \{j : g_j^\ell \neq 0\} . \tag{15.13}$$

Because the fronts must be disjoint sets, some of these nodes may be included already in previous fronts and must be eliminated as candidates for the new front. The new front, therefore, is the intersection of this set with the complement of the union of all previous fronts,

$$f^{\ell+1} = h^\ell \cap \left[\bigcup_{k=1}^\ell f^k \right]^c . \tag{15.14}$$

Starting from the first front,

$$f^1 = \{n_1^1\} , \tag{15.15}$$

the process continues until it reaches the null set.

15.3 The graph class

Listing 15.1 shows a truncated module containing the graph class. It follows a familiar design from previous examples and incorporates several familiar techniques. In particular, it contains an adjacency matrix as one of its data components. It is a straightforward modification of the compressed-sparse-row (CSR) matrix described in Chapter 7. Since the graph is undirected, its representation as a CSR matrix is the same as the alternative representation as a compressed-sparse-column (CSC) matrix.

Listing 15.1: The graph class.

```
module ClassGraph
  use ClassAdjacencyMatrix
  implicit none
  Type :: Graph
    integer :: me
    integer :: p
    integer :: nodes
    integer :: edges
    integer :: maxLocalEdges
    Type(AdjacencyMatrix)      :: Adjacency
    procedure(BFSInterface),   pass(G),pointer  &
                               :: bfs           => null()
    procedure(LtoGInterface),  pass(G),pointer  &
                               :: localToGlobal => null()
    procedure(GtoLInterface),  pass(G),pointer  &
                               :: globalToLocal => null()
  contains
    final :: deleteGraph
  end Type Graph

  interface Graph
    procedure newGraph
  end interface Graph
       :
end module ClassGraph
```

The constructor function reads the list of edges from a file that looks like the list of edges shown in tableau (15.7). It calls the constructor for the adjacency matrix and distributes the data across images using the same row-partitioned decomposition used for the CSR matrix. The constructor also fills in other information needed to describe the graph.

The graph class contains three procedure pointers, one for a breadth-first-search algorithm, one for a global-to-local index map, and one for a local-to-global index map. The programmer can associate these pointers with procedures different from the default procedures to investigate alternative ways to implement the search algorithm. The programmer can also add pointers for other graph algorithms.

15.4 A parallel breadth-first-search algorithm

Listing 15.2 shows an implementation of the breadth-first-search algorithm using the co-array model. It employs several familiar techniques from previous examples. A new feature of this algorithm is the use of lock variables described in Appendix A.9.

Listing 15.2: Parallel breadth-first-search code.

```fortran
function bfs(G,s) result(parent)
  use iso_fortran_env, only : lock_type
  use Collectives,     only : maxToAll
  implicit none
  Class(Graph),intent(in) :: G
  integer,intent(in)       :: s
  integer,allocatable      :: parent(:)
  Type(lock_type),allocatable :: parentLock[:]
  integer,allocatable          :: front(:)[:]
  integer,allocatable          :: prnt(:)[:]
  integer,allocatable          :: pos[:]
  integer :: f, i, j, k, L, len, m, n, owner
  integer :: OL(2)
    n = G%Adjacency%maxLocalRowDim
    allocate(parentLock[*])
    allocate(front(n)[*])
    allocate(prnt(n)[*])
    allocate(pos[*])
    allocate(parent(n))
    parent(:) = 0
    prnt(:)   = 0
    front(:)  = 0
    len       = 0
    OL = G%globalToLocal(s)
    owner = OL(1)
    i = OL(2)
    if(owner == G%me) then
      front(1) = s
      prnt(i)  = s
      len = 1
    end if
    do while (maxToAll(len) > 0)
      pos = len
      do f = 1,len
```

```
        OL = G%globalToLocal(front(f))
        owner = OL(1)
        j      = OL(2)
        do k = G%Adjacency%ia(j),G%Adjacency%ia(j+1)-1
            i = G%Adjacency%ja(k)
            OL = G%globalToLocal(i)
            owner = OL(1)
            L = OL(2)
            lock(parentLock[owner])
            if(prnt(L)[owner]==0) then
              prnt(L)[owner] = front(f)
              m = pos[owner] + 1
              pos[owner] = m
              front(m)[owner] = i
            end if
            unlock(parentLock[owner])
          end do
        end do
        sync all
        do f = len+1, pos
          front(f-len) = front(f)
        end do
        len = pos-len
      end do
      parent = prnt
end function bfs
```

The more familiar feature of the code is the use of maps between global and local node indices. The first step in the code determines ownership of the start node. The incoming argument is the global node index, and the function from global index to local index returns two result variables, the owner of the node and the local index of the node. The owner of the start node places it in its first front, sets the length of the front to one, and sets the parent array, at the local index, equal to the global index of the start node. The other images initialize their variables to zero.

The code then loops over fronts until all fronts are empty as determined by the max-to-all function from the collectives module. Each image independently examines the nodes in its front array, one by one, to find the edges connected to the node. The edges are obtained from the column-index array contained in the sparse adjacency matrix created by the graph constructor function. These edges are again labeled with global indices that are converted to the owner index and the local node index.

At this point in the algorithm, more than one image may want to determine whether an edge just found has already been entered into the parent array or whether it is new. Images must obtain the lock variable to guarantee that only one image at a time examines remote data structures. Since an image obtains the lock in different order from one run to the next, the parent array may be different from one run to the next. Nonetheless, the algorithm produces a valid parent array in each case.

The synchronization point guarantees that each image has examined all entries in its current front. Each image moves its new nodes forward in its front array, sets its length variable to the new value, which might be zero, and returns to the top of the loop to see if any other image has more nodes to examine. When every image has length zero, the code copies the local parent array to the output result array.

15.5 The Graph 500 benchmark

The breadth-first-search algorithm is the basis for the Graph 500 benchmark [97]. It mimics the older Top 500 benchmark [99] to reflect the increased interest in high-performance computers for non-numerical algorithms that dominate data analysis applications. A number of papers discuss parallel implementations of the algorithm [12] [24] [107].

The benchmark is an example of a weak-scaling experiment. Although the number of nodes in the graph must be large, between $n = 2^{26}$ for the toy problem and $n = 2^{42}$ for the huge problem, the number of processors used to run the algorithm is left undefined. The weak-scaling rule, therefore, is undefined and it is difficult to obtain any useful information of how one machine compares with another machine.

Yoo and co-workers, on the other hand, describe a specific weak-scaling experiment [107]. They pick a graph size $n(1)$ to run on one image and then scale the size as the image count increases,

$$n(p) = pn(1) , \tag{15.16}$$

such that the size on each image remains constant. Experimental measurements indicate that the execution time increases like the logarithm of the graph size,

$$t(p) = \log[\beta n(p)] . \tag{15.17}$$

Normalizing the time by the time on a single image,

$$\tau(p) = t(p)/t(1) , \tag{15.18}$$

yields the formula,

$$\tau(p) = 1 + (1/\gamma) \log p , \tag{15.19}$$

with

$$\gamma = \log[\beta n(1)] . \tag{15.20}$$

The function,

$$[\tau(p) - 1] \gamma = \log p , \tag{15.21}$$

therefore, should be independent of the implementation of the algorithm and independent of the machine used to execute the code.

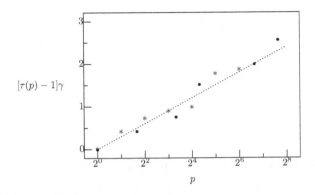

FIGURE 15.2: Weak-scaling results for the breadth-first-search algorithm. Bullets (•) mark the results from Yoo and co-workers [107]. Asterisks (*) mark results from the algorithm shown in Listing 15.2. The dotted line is the function $\log p$.

Figure 15.2 shows measured results using the code shown in Listing 15.2 compared with results from Yoo and co-workers. Although Yoo exerted much effort to implement a two-dimensional decomposition of the graph, compared with the one-dimensional decomposition used here, the weak-scaling results are the same when scaled appropriately. The two-dimensional decomposition corresponds to a block decomposition applied to sparse matrices similar to the block decomposition applied to dense matrices. A detailed exploration of such a decomposition is beyond the scope of this book. The interested reader may want to consult the literature [12] [24] [107].

The results shown in Figure 15.2 suggest that a single measurement on a single image is enough to characterize the breadth-first-search algorithm. The execution time on a single image,

$$t(1) = \log[\beta n(1)] \, , \tag{15.22}$$

differs from machine to machine just in the values of the initial problem size and the value of the scale parameter β. Other than that, two machines scale the same way with image count independent of the specific details of the implementation of the algorithm. The reported values for the absolute execution times for two different machines at different values of the image count and different problem sizes will, of course, be different. But without further information, there is no way to discover how different machines scale with image count and problem size.

15.6 Exercises

1. Prove that, for any node in the graph, no path to the start node is shorter than a path that follows an edge from one front f^ℓ to the next lower front $f^{\ell-1}$ until it reaches the start node.

2. What changes need to be made to the CSR matrix class to convert it to the adjacency matrix class?

3. Write a matrix-vector multiplication function for an adjacency matrix that takes as input a node index and returns a packed vector containing the nodes directly connected to the input node.

4. Modify the code shown in Listing 15.2 to generate a different, yet still valid, parent array.

5. Write a function that accepts a parent array as input and returns all the paths between the start node and all the other nodes in the graph. Return zero for a disconnected node.

6. Write a function that accepts a parent array and returns a level array listing the front containing each node in the graph. Return zero for a disconnected node.

Chapter 16

Epilogue

Writing code for scientific applications is hard; writing parallel code is even harder. The programmer can choose from several different languages and programming models depending on personal style and preference. Most of the hard parts of parallel programming, however, are independent of the language used or the programming model adopted. To a large extent, the constraints imposed by the physical problem and the numerical techniques used to solve the problem dictate the choices made for a parallel implementation.

This book has described the basic techniques used in parallel scientific applications independent of the language chosen or the programming model adopted. It developed the mathematics of particular algorithms, discussed data structures to support the algorithms, and displayed implementations of the algorithms. It used the Fortran language with co-arrays for these implementations because it provides a straightforward transcription of mathematics into code. The same algorithms could, of course, be implemented in other programming models using other data structures. But the underlying principles remain the same.

As hard as it is to write parallel code, it is even harder to write code that scales well and makes efficient use of ever-changing hardware. The book described performance modeling to set realistic expectations for scaling and performance. It gave precise definitions for often-used terms that describe performance models but are sometimes left ill-defined.

No language or programming model can solve the difficult problems encountered in writing parallel code without close attention by the programmer. The co-array programming model allows the programmer to concentrate more on the physical and numerical problem and less on the detailed syntax and symantics of the programming model. The co-array model closely matches the intuition of the Fortran programmer and maps rather closely onto the numerical algorithms being implemented.

Appendix A

A Brief Reference Manual for the Co-array Model

The co-array programming model is a standard feature of the Fortran language [56] [70]. Programmers write parallel application codes using co-array syntax to partition data structures and to move data between partitions. The co-array model is an extension of the long-standing rules of the language that allows the programmer to write parallel code using a natural Fortran syntax. Writing parallel algorithms within the co-array model requires minimal knowledge of the formal rules of syntax or semantics.

The co-array model is a Single-Program-Multiple-Data (SPMD) programming model with similarities and differences compared with other SPMD models. The co-array model is closely related to the model defined by the Message-Passing-Interface Library (MPI) [42] [91], to the Shmem model [20] [88], to the model defined by the Aggregate-Remote-Memory-Copy-Interface Library (ARMCI) [75] [76], and to the Global Arrays Library [74]. It is also closely related to the Bulk Synchronous Protocol (BSP) model [8], to the Unified Parallel C (UPC) model [15] [16] [102], and to the Titanium model [34] [106]. There is no inherent conflict between the co-array model and these other models. The programmer can, for example, insert co-array variables into an MPI program or insert MPI procedures into a co-array program without difficulty. The ability to replace one model with another model incrementally is an important design feature of the co-array model.

The co-array model is quite unlike the older High Performance Fortran (HPF) model [21] [49] [50] [60] [65] and quite unlike the X10 model from IBM [22] [23] or the Chapel model from Cray [17]. It is also quite unlike threaded models like the OpenMP model [19].

Based on an obvious extension to the Fortran language, the co-array model is carefully and deliberately designed to appeal to the intuition of the Fortran programmer. Versions of the idea, roughly called co-array models, have been proposed off and on over the years for the C and C++ languages [35] [93], even for the Python language [89]. How well the model fits these other languages is a subject for debate.

A.1   The image index

An image is a replication of a program. The programmer specifies the number of replications, and the run-time system assigns an image index to each replication. The number of images p and the image index q for a particular replication remain fixed during the execution of the program with the constraint,

$$1 \le q \le p \,, \tag{A.1}$$

beginning with image index one following the normal Fortran convention.

The run-time system assigns a physical processor to execute statements for each image and manages memory allocation for each image. It also provides a mechanism for a statement on one image to reference data that resides in memory assigned to another image and a mechanism for executing synchronization statements across images. The programmer knows nothing about how the system assigns image indices to physical processors. Particular implementations may provide ways, outside the co-array model, to specify this assignment.

During execution, the programmer can retrieve the number of images from the intrinsic function,

```
p = num_images ()
```

and the image index of the invoking image from the intrinsic function,

```
me = this_image ()
```

The simplest way to implement a compiler and a run-time system for the co-array model is to create a one-to-one correspondence between a single image and a single physical piece of hardware that executes instructions for that image.

But the co-array model does not require a one-to-one mapping. The programmer can insert directives into the code to spawn threads of execution for each image to share the work within a shared memory environment or to move work to an attached accelerator.

A particular implementation might allow a one-to-many model with one physical piece of hardware responsible for many images similar to a virtual-processor model like Charm++ [57] [58]. The run-time system might assign a set of images to each physical execution unit statically, or it might expect each physical execution unit to pick images dynamically from a list maintained by the system. This execution model may require extensive work in the run-time system to support it, but it may be an effective way to balance the workload across images, and it may be useful for developing and debugging programs on small systems with only a few physical processors.

A.2   Co-arrays and co-dimensions

A co-array variable is a variable declared with a co-dimension. For example, a real scalar variable,

```
real :: x
```

becomes a co-array variable by adding a co-dimension in square brackets,

```
real :: x[*]
```

The asterisk in the co-dimension declaration is treated the same way as the dimension declaration of an assumed-size array. Just as an assumed-size array inherits its size from its actual argument when a procedure is invoked, a co-array variable inherits its co-size from the run-time system at execution time with a value always equal to the number of images.

A co-array variable can be of type real, complex, integer, logical, character, or a variable of derived type. It has all the properties of a normal Fortran variable with the additional property that it is visible across images. The run-time system establishes a protocol that allows it to locate the address of a co-array variable in the local memory of every image. How this protocol works is specific to a particular implementation, and the programmer should make no assumptions about it.

A co-array variable may be a scalar or an array with one or more normal dimensions. For example, the declaration

```
complex :: c(m,n)[*]
```

defines a complex two-dimensional array that exists in the local memory assigned to each image and is visible across images. The normal dimensions correspond to the local size of the array assigned to each image. The size must be the same on each image to allow for simplification of the protocol for mapping addresses across images.

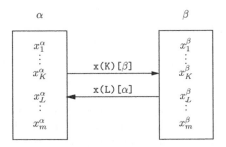

FIGURE A.1: A co-array variable assigned to local memory across images with the same size on each image. The run-time system establishes a protocol to execute statements for remote memory references. If the co-array is allocated symmetrically across images, the system can use the address in the local memory to construct the address in remote memory.

Co-dimensions are in a certain sense orthogonal to normal dimensions as suggested by Figure A.1. A normal dimension index designates an offset in local memory assigned to a particular image. A co-dimension index designates an offset across memory assigned to different images. In fact, the name co-array itself is meant to evoke an analogy with tensor algebra such that a normal array corresponds to a contravariant vector and a co-array corresponds to a covariant vector [79]. As indicated by the arrows in the figure, an image index in square brackets appended to a co-array variable is a reference to that variable assigned to another image.

Listing A.1: Reference to a co-array variable on another image.

```
integer :: m, me, alpha, beta
real    :: x(m)[*]
real    :: y(m)
   me = this_image()
      :
   if(me==alpha) then
      y(K)  = x(K)[beta]
   else if (me==beta) then
      y(L)  = x(L)[alpha]
   end if
```

Listing A.1 shows code that corresponds to the picture shown in Figure A.1. An important aspect of the co-array model is that a reference to a co-array variable without square brackets refers to the variable in the local memory assigned to the invoking image.

A.3 Relative co-dimension indices

The programmer can override the lower co-bound convention by declaring a co-array variable with a lower co-bound value different from one as shown in Listing A.2. The upper co-bound is still indicated by an asterisk, and the co-size always equals the number of images at run-time independent of the lower co-bound. Whatever the value of the lower co-bound, the programmer is responsible for generating valid co-dimension indices.

Listing A.2: Relative co-dimension index for a co-array variable with lower co-bound equal to L.

```
integer :: q(1)
real    :: x[L:*]
   q = this_image(x)
```

To help the programmer avoid errors, the co-array model provides another form of the intrinsic function that returns the image index relative to the declaration statement of a co-array variable. If the lower co-bound is L, as in Listing A.2, the function returns a value that obeys the constraint,

$$L \le q \le p + L - 1 , \tag{A.2}$$

If two co-array variables have different lower co-bounds, the co-dimension index returned by the intrinsic function is different for the two variables as shown in Listing A.3.

Listing A.3: Relative co-dimension indices for two co-array variables with different lower co-bounds.

```
integer :: meX(1), meY(1)
real    :: x[*]
real    :: y[0:*]
   meX = this_image(x)
   meY = this_image(y)
```

Another intrinsic function converts a relative image index for a particular co-array back to the absolute image index.

```
me = image_index(x,[q])
```

The second argument is an integer array of size equal to the number of co-dimensions of the first argument. If the lower co-bound is L and the value of q is valid, this function returns the image index $q - L + 1$. If the value of q is invalid, it returns zero. The two functions satisfy an inverse relationship.

```
zero = this_image() - image_index(x,this_image(x))
```

A.4   Co-array variables with multiple co-dimensions

The programmer can declare co-array variables with multiple co-dimensions, a common case being two co-dimensions,

```
real :: x[2,*]
```

although three or more are allowed. The programmer thinks of an image grid with a specified number of indices along the sides of the grid. The last co-dimension is an asterisk indicating that valid grid indices depend on the number of images at run-time.

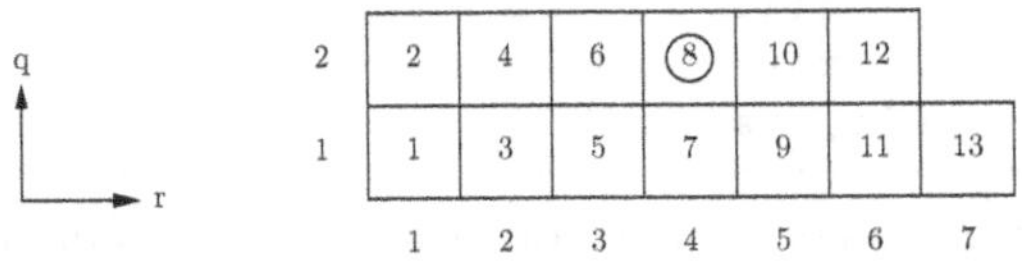

FIGURE A.2: The two-dimensional image grid for the co-array variable x[2,*]. The numbers inside the boxes are absolute image indices. The numbers along the sides of the grid are the co-dimension indices relative to the declaration statement for the co-array variable. Absolute image index 8 has grid indices [2,4].

Figure A.2 shows the case for a co-array variable declared with two valid image indices along the left side of the grid. When the number of images is thirteen at run-time, absolute image indices take values from one to thirteen as indicated by the numbers in each box. Image index eight, however, corresponds to grid indices [2,4]. If the first grid index is one, the second grid index can be as large as seven [1,7]. If the first grid index is two, however, the second index can be only as large as six [2,6]. Whether the grid indices increase up and to the right, as indicated by the arrows in Figure A.2, or down and to the left, or according to some other rule is purely a convention within the mind of the programmer.

Listing A.4: Co-dimension indices relative to a two-dimensional grid.

```
real      :: x[2,*]
integer :: me(2)
    me(:) = this_image(x)
    me(1) = this_image(x,1)
    me(2) = this_image(x,2)
```

The programmer can retrieve grid indices using the same intrinsic function as before. As shown in Listing A.4, the function returns an array of size two when there is no second argument or an array of size one from an alternative invocation with a second argument. The inverse function returns the absolute image index corresponding to an array argument that contains valid grid indices for the invoking image.

```
me = image_index(x,[q,r])
```

If either co-dimension index is out of bounds, the function returns the value zero. The rules for converting grid indices to an absolute image index are the same rules used for linearizing normal dimension indices. Normal dimension indices, however, imply contiguous data elements in local memory. Co-dimension indices imply nothing about how images are physically assigned across a machine.

The programmer may want to label the sides of the image grid starting, for example, with zero rather than one. Figure A.3 shows this case with the number of the absolute image index in each box. Image eight still has absolute index eight, but its grid indices [1,2] are different.

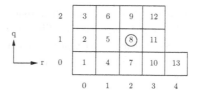

FIGURE A.3: Two-dimensional representation of images for a co-array variable declared as x[0:2,0:*]. The numbers inside the boxes are absolute image indices. The numbers along the sides of the grid are the grid indices based on the declaration statement. Image with index 8 has grid indices [1,2].

Co-array variables declared with two co-dimensions provide a convenient way to represent the relationship between nearest neighbors as shown in Figure A.4. Each image finds its own co-dimension indices within the grid [myP,myQ], and then associates its neighbors up-down-left-right by incrementing these indices. The relationship between neighbors is a logical convention in the programmer's mind usually representing the relationship between pieces of a physical domain decomposed into partitions. The up-down direction and the left-right direction are purely a convention that should be documented for each program. Otherwise, the convention used in any particular code may be difficult to determine.

A.5   Co-array variables of derived type

A variable of derived type can be declared as a co-array variable as long as it contains no components declared as co-array variables. Listing A.5 shows

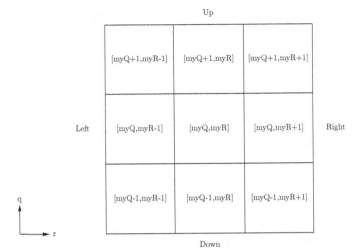

FIGURE A.4: Representation of a two-dimensional domain using co-dimensions for up-down-left-right directions. The programmer has adopted the convention that the up-direction corresponds to increasing values of the first co-dimension index and the right-direction corresponds to increasing values of the second co-dimension.

an example. Each image allocates the array component of the derived type variable independently with possibly different sizes at different locations in each local memory. Some images may even allocate the array with size zero or not allocate it at all. Data in the array component is visible across images indirectly through co-array syntax. The square bracket belongs to the co-array variable not to its array component.

Listing A.5: A co-array variable of derived type.

```
type A
   real,allocatable :: x(:)
end type A
   :
type(A)           :: vector[*]
real,allocatable  :: y(:)
   :
y = vector[alpha]%x(:)
```

Variables of derived type may contain components declared as co-array variables as long as the variable itself is not a co-array variable. Co-array syntax can appear only at one level of the derived type structure. No component of the derived type variable, declared itself as a co-array variable, can contain further co-array variable components. As shown in Listing A.6, the square bracket is now associated with the co-array component rather than with the variable itself.

Listing A.6: A variable of derived type with a component variable declared as a co-array variable.

```
type  B
   real,allocatable  ::  x(:)[:]
end  type  B
      :
type(B)              ::  vector
real,allocatable     ::  y(:)
      :
y = vector%x(:)[alpha]
```

Figure A.5 shows how reference to data on another image is a two-step process. To locate the array component in the local memory of another image, the invoking image must retrieve the remote address and the shape of the array as a first step and then retrieve data from the remote address as a second step. This double reference may result in degraded performance. Optimizing compilers, however, may generate code that pre-fetches addresses for remote references or may build a table of addresses generated at the time of allocation.

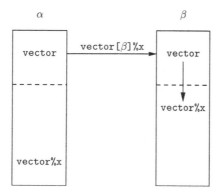

FIGURE A.5: A co-array variable of derived type. Image α first obtains the address of the array component from the derived type on the remote image. It then uses that address to perform a remote access of the data.

This double reference may be avoided when the co-array variable is a component of a normal variable of derived type. In this case, the roles of the variable and its component are reversed. The derived type variable resides at a different address on each image. The co-array component resides at the same local address on each image and is the same shape on each image.

A.6   Allocatable co-array variables

Allocatable co-array variables are declared with both the normal dimensions and the co-dimensions unresolved until run-time. The allocation statement must be located at the same line of code on each image with dimensions and co-dimensions the same on every image. The deallocation statement must also be located at the same line of code on each image. An allocation statement is a collective operation with hidden synchronization across all images before execution advances beyond the statement. A hidden synchronization also occurs before execution of a deallocation statement.

These hidden synchronization points are included to guarantee that a co-array variable has been allocated by all images before any communication between images takes place. Conversely, no co-array variable is deallocated while it may still be needed for communication. A co-array variable allocated within a procedure is automatically deallocated before exiting from the procedure unless the variable has the save attribute. The programmer must be aware of these hidden synchronization points to avoid deadlock.

A.7   Pointers

Co-array variables are not allowed to have the pointer attribute. A pointer assignment to a co-array variable is always a local assignment as shown in Listing A.7. Pointer assignment to an object on another image is illegal. The reason for this rule is that references to variables assigned to another image should always be apparent explicitly through the use of co-array syntax. Otherwise, remote references become implicit hidden references.

Listing A.7: Local association of a pointer with a co-array variable.

```
real,allocatable,target  ::  q[:]
real,pointer             ::  ptr
  allocate(q[*])
  ptr => q
```

Co-array variables of derived type may have pointer components. Such variables are useful for converting existing parallel code from another parallel model to the co-array model. Existing data structures typically cannot be declared directly as co-array variables because they involve different shapes across images. Converting them to co-array variables would require extensive changes to the code. Associating them with pointers inside co-array variables of derived type allows the programmer to maintain most of the existing code with minimal change and to replace communication constructs with co-array syntax [13].

A.8 Procedure interfaces

Co-dimensions are interpreted locally within each program unit. Co-dimension declarations can change from one procedure to another without changing how a co-array variable is assigned to memory across images. If a dummy argument is used for communication, it must be declared as a co-array variable, and the actual argument passed to the procedure must be a co-array variable. The programmer must supply an explicit procedure interface to prevent a normal variable being passed to a procedure expecting a co-array variable.

Co-array variables cannot be an automatic variable inside a procedure. If local variables are declared with explicit co-dimensions, they must have the save attribute. It is better, however, to declare local co-array variables as allocatable variables. The disadvantage of this approach is that it generates two implicit synchronization points inside the procedure, one when the variable is allocated and another when the variable is automatically deallocated at the end of the procedure. Since co-array variables are probably used for communication, some synchronization is required anyway so this penalty is usually not severe.

A.9 Execution control

The programmer inserts execution control statements into a code to divide statement execution into segments. Within each segment, statements execute independently for each image in the order they appear subject to the normal rules of the Fortran language. Control statements define a partial order of execution across images depending on where the programmer inserts the statements.

The co-array model deliberately places the burden of execution control with the programmer who knows better than the compiler or the run-time system where to insert control statements. Putting them in the right places, however, can be a difficult task, and mastering the art of parallel code development requires a great deal of practice and experience. The advantage of this approach is that compilers can use their normal optimization techniques within segments without worrying about execution order between images. A disadvantage is that the run-time system, at the end of a segment, may need to flush to memory all co-array variables held in local registers or caches and, at the start of a segment, it may need to wait for completion of all memory references to co-array variables and to discard any copies held in caches or registers.

The programmer decides where to insert an execution control statement based on the desired behavior of a program. If image r in segment σ_{j+1}^r needs the value of a variable defined by image q in segment σ_i^q, the programmer must insert an execution control statement to enforce the order relation,

$$\sigma_i^q \to \sigma_{j+1}^r \,, \tag{A.3}$$

where the arrow indicates that any statement executed in the segment on the left completes execution before any statement in the segment on the right begins execution [68]. The programmer must also guarantee that both images reach the corresponding control statement to enforce the order. Conversely, if image q in segment σ_{i+1}^q needs the value of a variable defined by image r in segment σ_j^r, the programmer must insert a control statement to enforce the order relation,

$$\sigma_j^r \to \sigma_{i+1}^q \,. \tag{A.4}$$

If the programmer fails to insert the control statement correctly, execution of the segments is unordered, and the program contains a memory race condition [73]. The program in that case may exhibit different behavior from one execution to another.

A.10   Full barriers

By inserting a full barrier into the code as shown in Listing A.8, the programmer imposes execution order across all images.

Listing A.8: A full barrier.

```
sync all
```

Before the statement executes, the system flushes co-array variables to memory and waits for remote memory activity to complete. If the number of barriers encountered differs from image to image, execution may stall in a state of deadlock. The compiler is not responsible for detecting deadlock situations, and the programmer must take care to avoid them.

Figure A.6 shows a program divided into three segments by two barriers. Some images may arrive at a barrier earlier than others. The run-time system may schedule an image out of execution to use its resources for other tasks. When the other images arrive at the barrier, they may proceed immediately

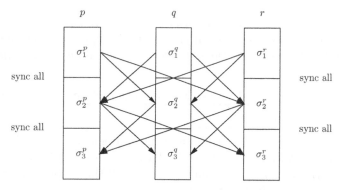

FIGURE A.6: A program divided into three segments by barriers. The arrows indicate the order relation between segments. All instructions at the tail of an arrow complete execution before any instruction at the head of the arrow begins execution.

because all the images have arrived. Some earlier images may need to be rescheduled before continuing execution through the barrier.

The co-array model, unlike some other programming models [8] [106], does not require barrier statements to match at the same point in the program for all images. A particularly difficult program to debug is one that contains multiple barriers that do not match correctly. The programmer intends each image to reach the same barrier in the same order but somehow writes code such that different images reach different barriers with the wrong match. Such code may reach a state of deadlock. The deadlock condition often occurs in a portion of the code far removed from the portion of code that actually causes the mismatch making the error difficult to detect and difficult to correct.

Even if the number of barrier statements executed for each image is the same, they may correspond with each other in some order different from the order intended by the programmer. The program continues execution, but the path of execution is wrong. This condition is referred to as livelock and is difficult to fix because the program continues execution with no indication of where the logic has gone wrong.

A.11 Partial barriers

To synchronize execution among just a subset of images, the programmer can insert a partial barrier statement as shown in Listing A.9.

Listing A.9: A partial barrier for images in a list.

```
sync images (list)
```

The argument in parentheses is an array containing a list of absolute image indices. An image in the list waits at the barrier until all other images in the list reach the barrier.

Listing A.10: A partial barrier with the list replaced by an asterisk.

```
sync images (*)
```

Listing A.10 shows an alternative form of the statement with the list replaced with an asterisk. An image reaching this barrier waits for all others to reach a matching barrier with a list containing its index. If this second form is invoked by all images, it is equivalent to a full barrier.

Listing A.11 shows a typical example that uses a partial barrier. All images with index greater than one wait for execution by image one to complete. Each image then passes the barrier without waiting for any of the other images. Image one waits until all other images have reached the barrier. The co-array syntax in the second segment passes a value from image one to the other images with the guarantee that they all obtain the same value.

Listing A.11: A program divided into two segments by a partial barrier.

```
real      :: x[*]
  if(me == 1) then
    x = y
    sync images(*)
  else
    sync images(1)
    y = x[1]
  end if
```

Figure A.7 shows how segment execution is ordered for this example. Notice how it differs from the picture shown in Figure A.6 for a full barrier.

Listing A.12: A forward cascade.

```
me = this_image()
p   = num_images()
left  = me-1
if(left == 0) left = p
right = me+1
if(right > p) right = 1
if(me == 1) then
  k = 1
  sync images(right)
  sync images(left)
else
  sync images(left)
  k = k[left] + 1
  sync images(right)
end if
```

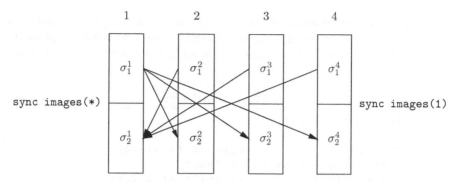

FIGURE A.7: Partial barriers for the code shown in Listing A.11. Execution by images two, three, and four waits for execution by image one. Execution continues independently across the barrier as suggested by the arrows that indicate segment order.

Listing A.12 shows how to create a cascade across images. Figure A.8 shows how the segments are ordered down the cascade.

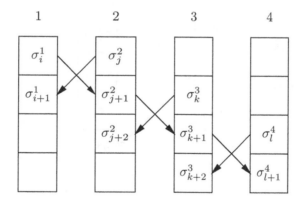

FIGURE A.8: Cascading statements for the code sample shown in Listing A.12. The arrows represent order relations between segments.

A.12   Critical segments and locks

Critical segments and lock variables provide another form of partial synchronization. Only one image executes within a critical or locked segment, but entrance into such segments is determined first come, first served. Listing A.13 shows an example of a critical segment. No other image control statement may be executed within a critical segment. Branching into or out of a critical segment is not permitted, and a critical segment must be properly nested with other constructs. Figure A.9 shows how segments are ordered for a critical segment.

Listing A.13: A critical segment.

```
integer :: total[*]
     :

   critical                ! begin critical segment
     total[q] = total[q] + count
   end critical            ! end critical segment
     :
```

If a program contains more than one critical construct, the system has no obligation to prevent two of them executing at the same time. For example, if two procedures contain the same construct shown in Listing A.13, the programmer must use some other form of synchronization to prevent them from executing simultaneously. Otherwise, a memory race condition may occur.

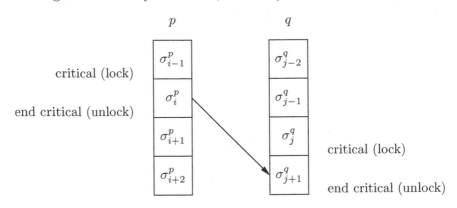

FIGURE A.9: Execution order for critical and locked segments.

Implementation of a critical segment with lock variables eliminates this kind of race condition. Listing A.14 shows the same example with the critical construct replaced by a lock construct. Entrance into the locked segment occurs only after obtaining the lock, an atomic operation allowing it to be obtained only one image at a time. Segment order is the same as for a critical segment as shown in Figure A.9.

Listing A.14: A locked segment.

```
use, intrinsic   :: iso_fortran_env
integer          :: total[*]
type(lock_type)  :: L[*]
     :

   lock(L[q])                ! begin locked segment
```

```
total[q] = total[q] + count
unlock(L[q])              ! end locked segment
```

Using critical segments or explicitly locked segments can be dangerous. In either case, errors in logic can result in bugs that are difficult to find.

A.13 Input/output

Each image has its own set of unit numbers that identify connected files. The code shown in Listing A.15, for example, opens the file named **data3** for image three and the file named **data4** for image four with the same local unit number. Since the two files are distinct, the two images obtain different data from different files even though they use the same local unit number.

Listing A.15: The open statement on different images.

```
character(len=*) :: header
        :
me = this_image()
if(me==3) open(10,file='data3')
if(me==4) open(10,file='data4')
if(me==3 .or. me==4) read(10,*) header
```

The intrinsic module **iso_fortran_env** contains integer constants for the default input unit, for the default output unit, and for the default error unit. The default input file is available for image one only. The code shown in Listing A.16, for example, employs image one to read data from standard input and then to broadcast it to the other images.

Listing A.16: Read on image one and broadcast to the others.

```
use, intrinsic :: iso_fortran_env
real :: z[*]
        :
me = this_image()
p  = num_images()
        :
if(me==1) then
    read(input_unit,*) z
    do q = 2,p
        z[q] = z
    end do
end if
sync all
```

Although the default unit for output is connected to a separate file on each image, the run-time system likely merges the sequences of output records into a single sequence of records. Each record is preserved with no mixing of records among images. The order of records from each image is preserved, but the order from image to image may vary from one system to another and even between runs on the same system. Likewise, files connected to the default error unit are separate for each image. It is expected, that the run-time system merges records into a single stream for the error unit or even into the same stream with all the records written to the default output unit.

The Fortan language provides no way to control the order of output from different images. The programmer may need to open a diagnostic file in append mode, one image at a time, write data to the file and then close the file retaining the data. This method is inefficient, but since it is likely used only for debugging purposes, efficiency is not important.

The current Fortran standard does not allow the same file to be connected to more than one image at a time.

A.14   Command line arguments

Command line arguments passed at run-time might specify, for example, co-dimensions used to partition a matrix and the matrix size as shown in Listing A.17.

Listing A.17: Passing command line arguments.

```
exec ./a.out p q m
```

Each image can retrieve these arguments with code like that shown in Listing A.18. The formatted internal read statements convert the character strings passed as arguments to integer values.

Listing A.18: Retrieving command line arguments.

```
n = command_argument_count ()

call get_command_argument (1, ch)
read( ch,  '(i10)' ) p

call get_command_argument (2, ch)
read( ch,  '(i10)' ) q
        :
call get_command_argument (n, ch)
read( ch,  '(i10)' ) m
```

A.15 Program termination

Provided no error conditions occur, each image continues execution until either it reaches a normal stop statement or it reaches the end of the program. If an error condition occurs, execution enters error termination rather than normal termination.

Listing A.19: Execution abort on all images.

```
error stop
```

In particular, execution of the statement shown in Listing A.19 aborts execution across all images. Propagation of error termination should be immediate within the performance limits of the hardware's ability to send signals across images, but the exact details depend on the implementation of the co-array model. The following list shows all causes for initiation of error termination.

- Execution of an `error stop` statement

- An error condition during execution of an `allocate`, `deallocate`, `sync all`, `sync images`, `lock`, `unlock`, or `sync memory` statement without a `stat=` specifier

- An error during execution of an `open`, `close`, `read`, `write`, `backspace`, `endfile`, `rewind`, `flush`, `wait`, or `inquire` statement without an `iostat=`, `end=`, or `err=` specifier

- An error during execution of a `print` statement

- An error during execution of the `execute_command_line` intrinsic subroutine and the optional `cmdstat` argument is not present

- An error condition outside Fortran

A.16 Inquiry functions

- `num_images()` returns the number of images at run-time as a default integer scalar.

- `lcobound(x[,dim,kind])` returns a lower co-bound or all lower co-bounds of a co-array variable.

  - `x` is a co-array variable of any type.

- If dim is absent, the result is a default integer array of rank one and size equal to the co-rank of the co-array variable; it holds all the lower co-bounds. If dim is present, it is a default integer scalar and the result is a default integer scalar holding the lower co-bound for co-subscript dim of the co-array variable.
- If kind is present, it is a scalar default integer specifying the kind of the result. If kind is absent, the result has the default kind.

- ucobound(x[,dim,kind]) is similar to lcobound except that it returns upper co-bounds.

A.17   Image index functions

- this_image() returns the index of the invoking image as a default integer scalar.

- this_image(x[,dim]) returns one or all of the co-subscripts that denote data on the invoking image.

    - x is a co-array variable of any type.
    - If dim is absent, the result is a default integer array of size equal to the co-rank of the co-array variable; it holds the set of co-subscripts of the co-array variable on the invoking image. If dim is present, it is a default integer scalar and the result is a default integer scalar holding co-subscript dim of the co-array variable on the invoking image.

- image_index(x,sub) returns the index of the image corresponding to the set of co-subscripts listed in the array sub for the co-array variable x as a default integer scalar.

    - x is a co-array variable of any type.
    - sub is a rank-one integer array of size equal to the co-rank of the co-array variable.

A.18   Execution control statements

Execution control statements divide the execution sequence on an image into segments. Many of the control statements have optional specifiers:

- stat=status where status is a scalar integer variable that is not otherwise involved in the statement. It is assigned the value zero if the statement is successful, the value stat_stopped_image, defined in the intrinsic module iso_fortran_env, if the statement encounters a stopped image, and a different value if it is otherwise unsuccessful.

- errmsg=message where message is a scalar default character variable with an error message assigned to it in the event of an error.

In the descriptions that follow, one or both of these specifiers form a list called stat-list.

- sync all(stat-list) Execution of a sync all statement synchronizes all the images. The n-th sync all statement to execute on image p corresponds to the n-th sync all statement to execute on image q. If the segments after corresponding sync all statements are S_i^p and S_j^q, the segments are ordered such that $S_{i-1}^p < S_j^q$ and $S_{j-1}^q < S_i^p$.

- allocate(allocation-list[,stat-list]) Execution of an allocate statement with an allocation-list that includes a co-array variable synchronizes all the images. The n-th execution of such an allocate statement on any image corresponds to the n-th execution of the same statement on any other image. No image may execute any statements following corresponding allocate statements until all images have completed the allocations.

- deallocate(object-list[,stat-list]) Execution of a deallocate statement with an object-list that includes a co-array variable synchronizes all the images. The n-th execution of such a deallocate statement on any image corresponds to the n-th execution of the same statement on any other image. No image is allowed to deallocate a co-array object until all images are ready to execute the corresponding deallocate statement.

- sync images(image-set[,stat-list]) Execution of a sync images statement on image p performs a synchronization of that image with each of the other images in the list image-set.

  - image-set is an integer scalar or array expression specifying one or more images or is * specifying all images.

Executions of sync images statements on images p and q correspond if the number of times image p has executed a sync images statement with q in its image set is the same as the number of times image q has executed a sync images statement with p in its image set. If the segments after corresponding sync images statements are S_i^p and S_j^q, the segments are ordered such that $S_{i-1}^p < S_j^q$ and $S_{j-1}^q < S_i^p$.

- `lock(lock-var[,acquired_lock=success][,stat-list])`

  - `lock-var` is a scalar variable of type `lock_type`. It is either a co-array variable or a subobject of a co-array variable.

  - `success` is a scalar logical variable.

  If `lock-var` is unlocked, successful execution of the statement on image p causes `lock-var` to become locked by image p. If `acquired_lock=success` is present, `success` is given the value true. If `lock-var` is locked by an image other than p, successful execution on image p of the statement with `acquired_lock=success` present causes `success` to be given the value false, or without `acquired_lock=success` present causes `lock-var` to become locked by image p after the other image has caused it to become unlocked. An error condition occurs if *lock-var* is already locked by the executing image. Any particular lock variable is successively locked and unlocked by a sequence of `lock` and `unlock` statements each separating two segments on the executing image. If execution of such an `unlock` statement on image p terminates segment S_{i-1}^{p} and is immediately followed in this sequence by execution of a `lock` statement on image q that starts segment S_{j}^{q}, then the segments are ordered such that $S_{i-1}^{p} < S_{j}^{q}$.

- `unlock(lock-var[,stat-list)`

  - `lock-var` is a scalar variable of type `lock_type`. It is either a co-array variable or a subobject of a co-array variable.

  If `lock-var` is locked by image p, it becomes unlocked after successful execution of the statement on image p. An error condition occurs if `lock-var` is not locked by the executing image. Any particular lock variable is successively locked and unlocked by a sequence of `lock` and `unlock` statements each separating two segments on the executing image. If execution of such an `unlock` statement on image p terminates segment S_{i-1}^{p} and is immediately followed in this sequence by execution of a `lock` statement on image q that starts segment S_{j}^{q}, then the segments are ordered such that $S_{i-1}^{p} < S_{j}^{q}$.

- **return, end** If an allocatable co-array variable is declared in a procedure, allocated in it, and remains allocated on return, it is automatically deallocated by the system. There is a hidden synchronization of all images. The statements that cause the return correspond.

- **block, end block** Execution of a statement that completes the execution of a `block` construct and results in an implicit deallocation of a co-array variable synchronizes all the images. The statements that complete the block correspond.

- `critical, end critical` The `critical` and `end critical` statements mark the start and end of a segment that is executed by a single image. If image q is the next image to execute the construct after image p, the `critical` statement on image q corresponds to the `end critical` statement on image p. If the segments in the construct on images p and q are S_i^p and S_j^q, then they are ordered such that $S_i^p < S_j^q$.

- `stop, end program` An image initiates normal termination if it executes a `stop` or `end program` statement. It may also be initiated during execution of a procedure defined by a C companion processor. The statements that initiate normal termination correspond and mark the end of the final segments executed by the images.

- `sync memory` The execution of a `sync memory` statement defines a boundary between two segments on the invoking image that can be ordered in some programmer-defined way with respect to segments on other images. The image executing this statement may not proceed beyond the statement until all memory traffic initiated by the image involving co-array variables has completed.

Bibliography

[1] Gene M. Amdahl. Validity of the single processor approach to achieving large-scale computing capabilities. In *AFIPS Spring Joint Computer Conference*, 1967.

[2] O. Axelsson and V.A. Barker. *Finite Element Solution of Boundary Value Problems*. SIAM, 2001.

[3] David H. Bailey. Twelve ways to fool the masses when giving performance results on parallel computers. *Supercomputing Review*, pages 54–55, August 1991.

[4] Pavan Balaji, Darius Buntinas, David Goodell, William Gropp, Sameer Kumar, Ewing Lusk, Rajeev Thakur, and Jesper Larsson Träff. Mpi on a million processors. In *Proceedings of the 16th European PVM/MPI Users' Group Meeting on Recent Advances in Parallel Virtual Machine and Message Passing Interface*, pages 20–30, Berlin, Heidelberg, 2009. Springer-Verlag.

[5] G. I. Barenblatt. *Scaling, Self-Similarity, and Intermediate Asymptotics*. Cambridge University Press, 1996.

[6] Pete Beckman, Kamil Iskra, Kazutomo Yoshii, Susan Coghlan, and Aroon Nataraj. Benchmarking the effects of operating system interference on extreme-scale parallel machines. *Cluster Computing*, 11(1):3–16, 2008.

[7] Garrett Birkhoff. *Hydrodynamics: A Study in Logic, Fact and Similitude*. Princeton University Press, 2nd edition, 1960.

[8] Rob H. Bisseling. *Parallel Scientific Computation: A Structured Approach using BSP and MPI*. Oxford University Press, 2004.

[9] G. Bosilca et al. Distributed Dense Numerical Linear Algebra Algorithms on Massively Parallel Architectures: DPLASMA. Technical Report UT-CS-10-660, Sept. 15, 2010, University of Tennessee Computer Science Technical Report, 2010.

[10] P. W. Bridgman. *Dimensional Analysis*. Yale University Press, New Haven, 2nd edition, 1931.

[11] Jehoshua Bruck, Ching-Tien Ho, Shlomo Kipnis, Eli Upfal, and Derrick Weathersby. Efficient Algorithms for All-to-All Communications in Multiport Message-Passing Systems. *IEEE Transactions on Parallel and Distributed Systems*, 8(11):1143–1156, November 1997.

[12] Aydin Buluç and Kamesh Madduri. Parallel breadth-first search on distributed memory systems. In *Proceedings of 2011 International Conference for High Performance Computing, Networking, Storage and Analysis*, SC '11, pages 65:1–65:12, New York, NY, USA, 2011. ACM.

[13] Paul M. Burton, Bob Carruthers, Gregory S. Fisher, Brian H. Johnson, and Robert W. Numrich. Converting the Halo-update subroutine in the Met Office unified model to Co-Array Fortran. In Walter Zwieflhofer and Norbert Kreitz, editors, *Developments in Teracomputing: Proceedings of the Ninth ECMWF Workshop on the Use of High Performance Computing in Meteorology, Reading, UK, November 13-17, 2000*, pages 177–188. World Scientific, 2001.

[14] David Callahan, John Cocke, and Ken Kennedy. Estimating interlock and improving balance for pipelined architectures. *Journal of Parallel and Distributed Computing*, 5:334–358, 1988.

[15] William W. Carlson and Jesse M. Draper. AC for the T3D. Technical Report SRC-TR-95-141, Supercomputer Research Center, Institute for Defense Analysis, 17100 Science Drive, Bowie, Maryland 20715-4300, February 1995.

[16] William W. Carlson, Jesse M. Draper, David E. Culler, Kathy Yelick, Eugene Brooks, and Karen Warren. Introduction to UPC Language Specification. Technical Report CCS-TR-99-157, IDA Center for Computing Sciences, Bowie, MD, 1999.

[17] Bradford L. Chamberlain, David Callahan, and Hans P. Zima. Parallel Programmability and the Chapel Language. *International Journal of High Performance Computing Applications*, 21(3):291–312, August 2007.

[18] Ernie Chan, Marcel Heimlich, Avi Purkayastha, and Robert van de Geijn. Collective communication: theory, practice, and experience. *Concurrency and Computation: Practice and Experience*, 19(13):1749–1783, 2007.

[19] Rohit Chandra, Ramesh Menon, Leo Dagum, David Kohr, Dror Maydan, and Jeff McDonald. *Parallel Programming in OpenMP*. Morgan-Kaufmann, Inc., 2001.

[20] Barbara Chapman, Tony Curtis, Swaroop Pophale, Chuck Koelbel, Jeff Kuehn, Stephen Poole, and Lauren Smith. Introducing OpenSHMEM. In *Proceedings of PGAS '10*. New York, New York, October 12-15, 2010.

[21] Barbara Chapman, Piyush Mehrotra, and Hans Zima. Programming in Vienna Fortran. *Scientific Programming*, 1:31–50, 1992.

[22] Philippe Charles, Christopher Donawa, Kemal Ebcioglu, Christian Grothoff, Allan Kielstra, Christoph von Praun, Vijay Saraswat, and Vivek Sarkar. X10: An Object-Oriented Approach to Non-Uniform Cluster Computing. In *Proceedings of the ACM 2005 OOPSLA Conference*, October 2005.

[23] Philippe Charles, Christian Grothoff, Vijay Saraswat, Christopher Donawa, Allan Kielstra, Kemal Ebcioglu, Christoph von Praun, and Vivek Sarkar. X10: an object-oriented approach to non-uniform cluster computing. In *OOPSLA '05: Proceedings of the 20th annual ACM SIGPLAN conference on Object-oriented programming, systems, languages, and applications*, pages 519–538, New York, NY, USA, 2005. ACM.

[24] Fabio Checconi, Fabrizio Petrini, Jeremiah Willcock, Andrew Lumsdaine, Anamitra Choudhury, and Yogish Sabharwal. Breaking the speed and scalability barriers for graph exploration on distributed-memory machines. In *Proceedings SC12*, 10-16 November, Salt Lake City 2012.

[25] C. Chevalier and F. Pellegrini. PT-SCOTCH: A tool for efficient parallel graph ordering. *Parallel Computing*, 34:318–331, 2008.

[26] Jaeyoung Choi, Jack J. Dongarra, and David W. Walker. Parallel matrix transpose algorithms on distributed memory concurrent computers. *Parallel Computing*, 21:1387–1405, 1995.

[27] François Cuvelier. Méthodes des éléments finis. De la théorie á la programmation. http://www.math.univ-paris13.fr/~cuvelier/docs/poly/polyFEM2D.pdf, January 2013.

[28] François Cuvelier, Caroline Japhet, and Gilles Scarella. An efficient way to perform the assembly of finite element matrices in Matlab and Octave. Technical Report 8305, INRIA Paris-Rocquencourt, Project Teams Pomdapi, May 2013.

[29] James W. Demmel, Nicholas J. Higham, and Robert S. Schreiber. Stability of Block *LU* Factorization. *Numerical Linear Algebra with Applications*, 2(2):173–190, 1995.

[30] Jack Dongarra, Michael A. Heroux, and Piotr Luszczek. HPCG Benchmark: a New Metric for Ranking High Performance Computing Systems. Technical Report UT-EECS-15-736, Electrical Engineering and Computer Science Department, Knoxville, Tennessee, November 2015.

[31] Jack Dongarra and Julien Langou. The Problem with the Linpack Benchmark 1.0 Matrix Generator. Technical report, http://arxiv.org/pdf/0806.4907.pdf, 2008.

[32] Jack J. Dongarra, Robert A. van de Geijn, and David W. Walker. Scalability issues affecting the design of a dense linear algebra library. *Journal of Parallel and Distributed Computing*, 22:523–537, 1994.

[33] Jack J. Dongarra and David W. Walker. Software libraries for linear algebra computations on high performance computers. *SIAM Review*, 37(2):151–180, June 1995.

[34] Tarek El-Ghazawi, William Carlson, Thomas Sterling, and Katherine Yelick. *UPC: Distributed Shared Memory Programming*. John Wiley and Sons, Inc., Hoboken, New Jersey, 2005.

[35] M. Eleftheriou, S. Chatterjee, and J. E. Moreira. A C++ implementation of the co-array programming model for Blue Gene/L. In *Proceedings 16th International Parallel and Distributed Processing Symposium (IPDPS 2002)*, 15-19 April 2002.

[36] M. Faverge et al. Designing LU-QR hybrid solvers for performance and stability. In *IPDPS, Phoenix, AZ, 2014*. IEEE Computer Science Press, 2014.

[37] G. C. Fox, S. W. Otto, and A. J. G. Hey. Matrix algorithms on a hypercube I: Matrix multiplication. *Parallel Computing*, 4:17–31, 1987.

[38] Matteo Frigo and Steven G. Johnson. The Design and Implementation of FFTW3. *Proceedings of the IEEE*, 93(2):216–231, February 2005.

[39] Erich Gamma, Richard Helm, Ralph Johnson, and John Vlissides. *Design Patterns: Elements of Reusable Object-Oriented Software*. Addison-Wesley, 1995.

[40] David Gottlieb and Steven A. Orszag. *Numerical Analysis of Spectral Methods: Theory and Applications*. SIAM, 1977.

[41] Bruce Greer and Greg Henry. High Performance Software on Intel Pentium Pro Processors or Micro-Ops to TeraFLOPS. In *Proceedings of Supercomputing '97*, pages 1–13, 1997.

[42] William Gropp, Ewing Lusk, and Anthony Skjellum. *Using MPI, Portable Parallel Programming with the Message-Passing Interface*. The MIT Press, 1994.

[43] John Gustafson, Diane Rover, Stephen Elbert, and Michael Carter. The design of a scalable, fixed-time computer benchmark. *Journal of Parallel and Distributed Computing*, 12(4):388–401, August 1991.

[44] John L. Gustafson. Amdahl's Law Reevaluated. *Communications of the ACM*, 31(5):532–533, May 1988.

[45] John L. Gustafson. The Scaled-Sized Model: A Revision of Amdahl's Law. In L. P. Kartashev and S. I. Kartashev, editors, *Proceedings of the Third International Conference on Supercomputing: Supercomputing '88*, volume 2, pages 130–133, St. Petersburg, FL, 1988.

[46] John L. Gustafson, Gary R. Montry, and Robert E. Benner. Development of Parallel Methods for a 1024-Processor Hypercube. *SIAM Journal on Scientific and Statistical Computing*, 9(4):609–638, July 1988.

[47] Stephen Hawking. *God Created the Integers*. Running Press, 2005.

[48] Bruce Hendrickson and Robert Leland. An improved spectral graph partitioning algorithm for mapping parallel computations. *SIAM Journal on Scientific Computing*, 16(2):452–469, March 1995.

[49] High Performance Fortran Forum. High Performance Fortran/Journal of Development. *Scientific Programming*, 2(1-2), 1993.

[50] High Performance Fortran Forum. HPF-2 scope of activities and motivating applications. Technical Report CRPC-TR94492, Center for Research on Parallel Computation, Rice University, Houston, TX, November 1994.

[51] R. W. Hockney and I. J. Curington. f-half: a Parameter to Characterise Memory and Communication Bottlenecks. *Parallel Computing*, 10:277–286, 1989.

[52] Roger W. Hockney. *The Science of Computer Benchmarking*. Society for Industrial and Applied Mathematics (SIAM), Philadelphia, 1996.

[53] Torsten Hoefler and Roberto Belli. Scientific benchmarking of parallel computing systems: Twelve ways to tell the masses when reporting performance results. In *Proceedings of the International Conference for High Performance Computing, Networking, Storage and Analysis*, SC '15, pages 73:1–73:12, New York, NY, USA, 2015. ACM.

[54] Torsten Hoefler, Timo Schneider, and Andrew Lumsdaine. The impact of network noise at large-scale communication performance. In *IEEE International Symposium on Parallel & Distributed Processing (IPDPS 2009)*, pages 1–8, May 2009.

[55] Niclas Jansson. Optimizing sparse matrix assembly in finite element solvers with one-sided communication. In Michel Daydé, Osni Marques, and Kengo Nakajima, editors, *High Performance Computing for Computational Science - VECPAR 2012*, volume 7851 of *Lecture Notes in Computer Science*, pages 128–139. Springer Berlin Heidelberg, 2013.

[56] ISO/IEC JTC1/SC22. International Standard ISO/IEC 1539-1:2010(E) Information technology - Programming languages - Fortran - Part 1: Base language. *ISO/IEC*, Geneva, 2010.

[57] L. V. Kale and Sanjeev Krishnan. Charm++: Parallel Programming with Message-Driven Objects. In Gregory V. Wilson and Paul Lu, editors, *Parallel Programming using C++*, pages 175–213. MIT Press, 1996.

[58] Laxmikant Kale. Charm++. In D. Padua, editor, *Encyclopedia of Parallel Computing*. Springer Verlag, 2011.

[59] George Karypis and Vipin Kumar. A fast and high quality multilevel scheme for partitioning irregular graphs. In *International Conference on Parallel Processing*, pages 113–122, 1995.

[60] Ken Kennedy, Charles Koelbel, and Hans Zima. The rise and fall of High Performance Fortran: an historical object lesson. In *HOPL III: Proceedings of the third ACM SIGPLAN conference on History of programming languages*, pages 7-1-7-22, New York, NY, USA, 2007. ACM Press.

[61] D. J. Kerbyson, H. J. Alme, A. Hoisie, F. Petrini, H. J. Wasserman, and M. Gittings. Predictive Performance and Scalability Modeling of a Large-scale Application. In *Proceedings of Supercomputing 2001*, Denver, CO, November 2001.

[62] Darren J. Kerbyson, Adolfy Hoisie, Scott Pakin, Fabrizio Petrini, and Harvey J. Wasserman. A performance evaluation of an Alpha EV7 processing node. *The International Journal of High Performance Computing Applications*, 18(2):199–209, May 1, 2004.

[63] Darren J. Kerbyson, Adolfy Hoisie, and Harvey J. Wasserman. Modeling the Performance of Large-Scale Systems. *IEE Proceedings: Software, Inst. Electrical Engineers*, 150(4), July 2003.

[64] Darren J. Kerbyson, Harvey J. Wasserman, and Adolfy Hoisie. Exploring advanced architectures using performance prediction. In *Proceedings of the International Workshop on Innovative Architecture for Future Generation High-Performance Processors and Systems (IWIA'02)*, page 27, Washington, DC, USA, 2002. IEEE Computer Society.

[65] Charles Koelbel, D. Loveman, R Schreiber, G. Steele, and M. Zosel. *The High Performance Fortran Handbook*. The MIT Press, Cambridge, MA, 1994.

[66] Rahul Kumar, Amith Mamidala, and D. K. Panda. Scaling alltoall collective on multi-core systems. In *IEEE International Symposium on Parallel and Distributed Processing (IPDPS)*, pages 1–8, 14-18 April 2008.

[67] V. Kumar, A.Y. Grama, and N.R. Vempaty. Scalable load balancing techniques for parallel computers. *Journal of Parallel and Distributed Computing*, 22(1):60–79, July 1994.

[68] Leslie Lamport. A new approach to proving the correctness of multiprocessor programs. *ACM Transactions on Programming Languages and Systems*, 1(1):84–97, July 1979.

[69] John D. McCalpin. Memory bandwidth and machine balance in current high performance computers. *IEEE Computer Society; Technical committee on Computer Architecture Newsletter*, December 1995.

[70] Michael Metcalf, John Reid, and Malcolm Cohen. *Modern Fortran Explained*. Oxford University Press, 2011.

[71] Douglas Miles. Compute intensity and the FFT. *Proceedings Supercomputing 1993*, pages 676–684, 1993.

[72] K. W. Morton and D. F. Mayers. *Numerical Solution of Partial Differential Equations*. Cambridge University Press, 2nd edition, 2005.

[73] Robert H. B. Netzer and Barton P. Miller. What are race conditions?: Some issues and formalizations. *ACM Letters on Programming Languages and Systems (LOPLAS)*, 1(1):74–88, March 1992.

[74] J. Nieplocha, R.J. Harrison, and R.J. Littlefield. Global Arrays: A portable shared memory model for distributed memory computers. In *Proceedings of Supercomputing '94*, pages 340–349, 1994.

[75] J. Nieplocha, J. Ju, Vinod Tipparaju, and E. Apra. One-sided communication on clusters with myrinet. *Cluster Computing*, 6:115–124, 2003.

[76] Jarek Nieplocha and B. Carpenter. ARMCI: A portable remote memory copy library for distributed array libraries and compiler run-time systems. In *Lecture Notes in Computer Science*, volume 1586. Springer-Verlag, 1999.

[77] Robert W. Numrich. Computational Force, Mass, and Energy. *International Journal of Modern Physics C*, 8(3):437–457, June 1997.

[78] Robert W. Numrich. F^{--}: A parallel extension to Cray Fortran. *Scientific Programming*, 6(3):275–284, Fall 1997.

[79] Robert W. Numrich. Parallel numerical algorithms based on tensor notation and Co-Array Fortran syntax. *Parallel Computing*, 31:588–607, 2005.

[80] Robert W. Numrich. A note on scaling the Linpack benchmark. *Journal of Parallel and Distributed Computing*, 67(4):491–498, April 2007.

[81] Robert W. Numrich. Computational forces in the Linpack benchmark. *Journal of Parallel and Distributed Computing*, 68(9):1283–1290, September 2008.

[82] Robert W. Numrich. Computational forces in the SAGE benchmark. *Journal of Parallel and Distributed Computing*, 69(3):315–325, 2009.

[83] Robert W. Numrich. Computer performance analysis and the Pi Theorem. *Computer Science - Research and Development*, 29(1):45–71, 2014.

[84] Robert W. Numrich and Michael A. Heroux. Self-similarity of parallel machines. *Parallel Computing*, 37(2):69–84, February 2011.

[85] Fabrizio Petrini, Darren J. Kerbyson, and Scott Pakin. The case of the missing supercomputer performance: Achieving optimal performance on the 8,192 processors of ASCI Q. In *Proceedings of SC2003*, Phoenix, Arizona, November 15–21, 2003.

[86] Steve Plimpton. Fast parallel algorithms for short-range molecular dynamics. *Journal of Computational Physics*, 117:1–19, 1995.

[87] Steve Plimpton, Roy Pollock, and Mark Stevens. Particle-mesh ewald and rrespa for parallel molecular dynamics. In *Proceedings of the Eighth SIAM Conference on Parallel Processing for Scientific Computing*, 1997.

[88] Quadrics. *Shmem Programming Manual*. Quadrics Supercomputing World Ltd, 2001.

[89] Craig E. Rasmussen, Matthew J. Sottile, Jarek Nieplocha, Robert W. Numrich, and Eric Jones. Co-Array Python: A Parallel Extension to the Python Language. In *Proceedings of Euro-Par 2004 Parallel Processing: 10th International Euro-Par Conference*, pages 632–637. Lecture Notes in Computer Science Number 3149, Springer-Verlag GmbH, Pisa, Italy, August 31-September 3, 2004.

[90] Yousef Saad. *Iterative Methods for Sparse Linear Systems*. SIAM, 2nd edition, 2003.

[91] Marc Snir and William Gropp. *MPI: The Complete Reference*. MIT Press, 1998.

[92] Gilbert Strang and George Fix. *An Analysis of the Finite Element Method*. Wellesley-Cambridge Press, 2nd edition, 2008.

[93] Antoine Tran Tan and Hartmut Kaiser. Extending C++ with Co-array Semantics. In *Proceedings of the 3rd ACM SIGPLAN International Workshop on Libraries, Languages, and Compilers for Array Programming*, ARRAY 2016, pages 63–68, New York, NY, USA, 2016. ACM.

[94] Clive Temperton. Self-sorting in-place Fast Fourier Transforms. *SIAM Journal on Scientific and Statistical Computing*, 12(4):808–823, July 1991.

[95] Clive Temperton. A generalized prime factor FFT algorithm for any $N = 2^p 3^q 5^r$. *SIAM Journal on Scientific and Statistical Computing*, 13(3):676–686, May 1992.

[96] Rajeev Thakur, Rolf Rabenseifner, and William Gropp. Optimization of Collective Communication Operations in MPICH. *International Journal of High Performance Computing Applications*, 19(1):49–66, 2005.

[97] The Graph 500 List. `http://www.graph500.org/`, accessed 31 July 2016.

[98] The HPCG Benchmark. `http://www.hpcg-benchmark.org`, accessed 23 April 2018.

[99] Top 500 Benchmark. http://www.top500.org/lists/linpack.php.

[100] Lloyd N. Trefethen. *Spectral Methods in MatLab*. SIAM, 2000.

[101] Lloyd N. Trefethen and Robert S. Schreiber. Average-case stability of Gaussian elimination. *SIAM J. Matrix Analysis*, 11(3):335–360, July 1990.

[102] UPC Consortium. UPC Language Specifications V1.2. www.gwu.edu/ upc/publications/LBNL-59208.pdf, May 31 2005.

[103] Nisheeth K. Vishnoi. The impact of noise on the scaling of collectives: the nearest neighbor model. In *Proceedings of the 14th international conference on High performance computing*, HiPC'07, pages 476–487, Berlin, Heidelberg, 2007. Springer-Verlag.

[104] Wenqing Wang and Olaf Kolditz. Sparse Matrix and Solver Objects for Parallel Finite Element Simulation of Multi-field Problems. In *High Performance Computing and Applications, Lecture Notes in Computer Science*, volume 5938, pages 418–425. Springer, 2010.

[105] Patrick J. Worley. The effect of time constraints on scaled speedup. *SIAM J. Sci. Stat. Comput.*, 11(5):838–858, September 1990.

[106] Kathy Yelick, Luigi Semenzato, Geoff Pike, Carleton Miyamoto, Ben Liblit, Arvind Krishnamurthy, Paul Hilfinger, Susan Graham, David Gay, Phil Colella, and Alex Aiken. Titanium: a high-performance Java dialect. *Concurrency: Practice and Experience*, 10(11-13):825–836, September-November 1998.

[107] Andy Yoo, Edmond Chow, Keith Henderson, William McLendon, Bruce Hendrickson, and Umit Catalyurek. A scalable distributed parallel breadth-first search algorithm on BlueGene/L. In *Proceedings SC05 (Lawrence Livermore National Laboratory, Technical Report UCRL-CONF-213752, 20 July 2005)*, 12 November, 2005.

Index

Printed and bound by CPI Group (UK) Ltd, Croydon, CR0 4YY

23/10/2024

01777697-0002